Developing Numeracy
MENTAL MATHS

ACTIVITIES FOR THE DAILY MATHS LESSON

year
6

Hilary Koll and Steve Mills

A & C BLACK

Contents

Published 2004 by A & C Black Publishers Limited
37 Soho Square, London W1D 3QZ
www.acblack.com

ISBN 0-7136-6915-2

Copyright text © Hilary Koll and Steve Mills, 2004
Copyright illustrations © David Benham, 2004
Copyright cover illustration © Charlotte Hard, 2004
Editors: Lynne Williamson and Marie Lister

The authors and publishers would like to thank Jane McNeill and Catherine Yemm for their advice in producing this series of books.

A CIP catalogue record for this book is available from the British Library.

Printed and bound in Great Britain by Cromwell Press Ltd, Trowbridge.

A & C Black uses paper produced with elemental chlorine-free pulp, harvested from managed sustainable forests.

Introduction

Developing Numeracy: Mental Maths is a series of seven photocopiable activity books designed to be used during the daily maths lesson. This book focuses on the skills and concepts for mental maths outlined in the National Numeracy Strategy *Framework for teaching mathematics* for Year 6. The activities are intended to be used in the time allocated to pupil activities; they aim to reinforce the knowledge and develop the facts, skills and understanding explored during the main part of the lesson. They provide practice and consolidation of the objectives contained in the framework document.

Mental Maths Year 6

To calculate mentally with confidence, it is necessary to understand the three main aspects of numeracy shown in the diagram below. These underpin the teaching of specific mental calculation strategies.

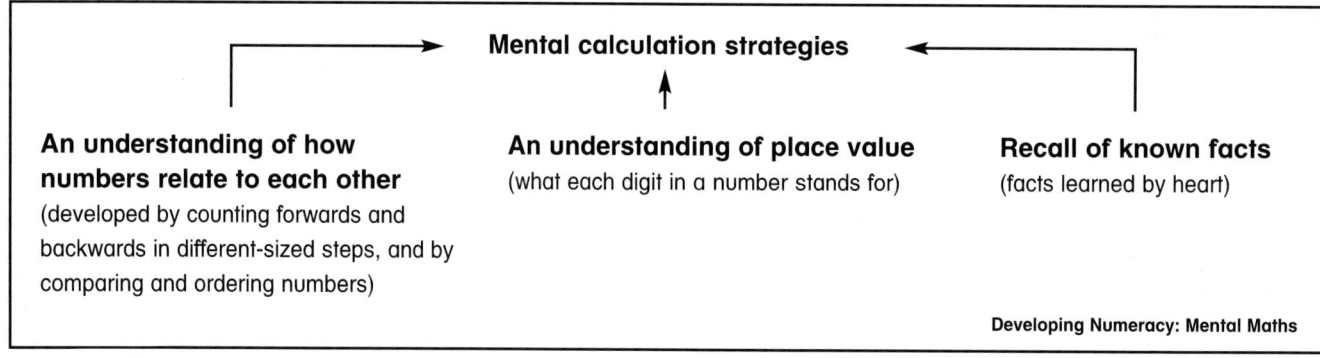

Mental calculation strategies

An understanding of how numbers relate to each other
(developed by counting forwards and backwards in different-sized steps, and by comparing and ordering numbers)

An understanding of place value
(what each digit in a number stands for)

Recall of known facts
(facts learned by heart)

Developing Numeracy: Mental Maths

Year 6 supports the teaching of mental maths by providing a series of activities which develop these essential skills. On the whole the activities are designed for children to work on independently, although this is not always possible and occasionally some children may need support.

Year 6 develops concepts and skills for the different aspects of numeracy in the following ways:

An understanding of how numbers relate to each other
- recognising and extending number sequences, such as the sequence of square numbers, or the sequence of triangular numbers 1, 3, 6, 10, 15…;
- counting on in steps of 0·1, 0·2, 0·25, 0·5…, and then back;
- recognising multiples up to 10 × 10, and squares of numbers to at least 12 × 12;
- knowing and applying simple tests of divisibility, and finding simple common multiples;
- factorising numbers to 100 into prime factors;
- using negative numbers and finding the difference between a positive and a negative integer, or two negative integers.

An understanding of place value
- multiplying and dividing decimals mentally by 10 or 100, and integers by 1000, and explaining the effect;
- knowing what each digit represents in a number with up to three decimal places;
- partitioning when multiplying.

Recall of known facts
Know by heart or derive quickly:
- decimals that total 1 or 10, all two-digit pairs that total 100, and all pairs of multiples of 10 with a total of 1000;
- multiplication facts up to 10 × 10, and squares of multiples of 10 to 100;
- division facts corresponding to tables up to 10 × 10;
- doubles of two-digit numbers, doubles of multiples of 10 to 1000, doubles of multiples of 100 to 10 000, and the corresponding halves.

Mental calculation strategies
- finding a difference by counting up;
- adding or subtracting the nearest multiple of 10, 100 or 1000, then adjusting;
- adding several numbers and adding three or four multiples of 10;
- adding sets of numbers by multiplying and adjusting;
- using known number facts and place value to consolidate mental addition and subtraction;
- using factors, related facts and doubling or halving;
- partitioning when multiplying;
- using brackets;
- using known number facts and place value to consolidate mental multiplication and division;
- finding simple percentages of small whole-number quantities;
- identifying and using appropriate operations to solve word problems (converting pounds to euros, and vice versa).

Extension

Many of the activity sheets end with a challenge (**Now try this!**) which reinforces and extends the children's learning, and provides the teacher with an opportunity for assessment. On occasion it may be helpful to read the instructions with the children before they begin the activity. For some of the challenges the children will need to record their answers on a separate piece of paper.

Organisation

Very little equipment is needed, but it will be useful to have the following resources available: coloured pencils, counters, dice, scissors, number lines and number tracks.

To help teachers select appropriate learning experiences for the children, the activities are grouped into sections within the book. However, the activities are not expected to be used in this order; the sheets are intended to support, rather than direct, the teacher's planning.

Some activities can be made easier or more challenging by masking or substituting numbers. You may wish to re-use some pages by copying them onto card and laminating them.

Teachers' notes

Brief notes are provided at the foot of each page giving ideas and suggestions for maximising the effectiveness of the activity sheets. These can be masked before copying.

Whole-class warm-up activities

The following activities provide some practical ideas which can be used to introduce the main teaching part of the lesson.

Crowns

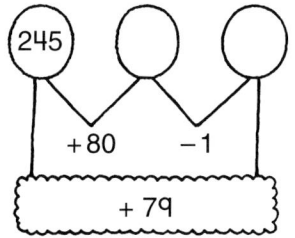

Draw a crown to demonstrate the strategy of adding or subtracting the nearest multiple of 10, 100 or 1000, then adjusting. Use an example such as 245 + 79. Write '245', '+ 79', '+ 80' and '– 1' on the crown as shown. With your finger, trace the route from 245 past + 80 and write 325 in the middle circle. Then trace past – 1 and write 324 in the last circle. Show that this is the answer to 245 + 79. Give other additions and subtractions and ask the children to draw their own crowns to find the answers.

Function machines

Draw a simple machine on the board and write a single-step or multi-step operation inside it, such as '– 29' or '+ 17 × 3'. Say a number, such as 20, and ask for the output number after the rule has been applied.

Function machines can also be used in the following ways:
- give the output number and ask the children for the input number that produces it;
- give pairs of input and output numbers that correspond to the same rule, and ask the children to work out the rule. For example, for the rule '× 7 – 4', you could give the following pairs: 3 and 17, 5 and 31, 7 and 45. Rather than revealing the rule to the rest of the class when they have worked it out, the children could suggest further pairs of input and output numbers that correspond to the rule.

Alphabet sums

Write the letters of the alphabet with the numbers 0·1 to 2·6 beneath. Explain how to use the letter values to find the total of a word: for example, 'BAG' has a value of 1 (the total of 0·2 + 0·1 + 0·7). Ask: *What is your first name worth? Who can find the four-letter word with the highest total? Can anyone find a four-letter word with a total of 1·2?* (BEAD = 0·2 + 0·5 + 0·1 + 0·4).

Totals

Ask the children to draw four boxes in a row:

Explain that you are going to call out four single-digit numbers, each of which should be written in one of the boxes. The children must decide where to write each number before the next one is called. The target can be to make the largest number, smallest number, number nearest to 5000, and so on.

Bingo

Ask the children to draw a 4 × 3 grid and to write a number between 20 and 50 in each of the 12 boxes. (Explain that all the numbers must be different.) Call out questions with answers that lie between 20 and 50: for example, *6 times 4* or *70 minus 29*. Any children who have the answers in their grids may cross them out. The winner is the first player to cross out all 12 numbers. Keep a record of the questions asked, so that you can check the winner's grid and see which questions cause difficulty.

Money bags

These bags contain 10p coins.

- Look at the total on each bag. Work out
  how many 10p coins there are.

 **1.** £1.20

 **2.** £2.50

 **3.** £33.40

 **4.** £7.80

5. £80.00

_12_

 **6.** £14.00

 **7.** £12.50

 **8.** £437.80

 **9.** £300.50

 **10.** £488.00

These bags contain 1p coins.

- Work out how many 1p coins there are.

 **11.** £50.50

 **12.** £22.80

 **13.** £6.30

 **14.** £90.90

 **15.** £18.09

 **16.** £123.45

 **17.** £1.25

 **18.** £66.40

 **19.** £9.05

 **20.** £4.85

 **Now try this!**

- Write the total amount on each bag.

$20 \times £10$
$30 \times £100$
 £

$14 \times £10$
$40 \times £100$
 £

$142 \times £10$
$150 \times £100$
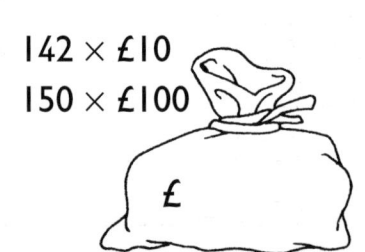 £

Teachers' note Discuss with the children that these questions can be answered by multiplying each decimal by 10 (questions 1 to 10) or by 100 (questions 11 to 20).

**Developing Numeracy
Mental Maths Year 6
© A & C BLACK**

Interactive quiz

• **Shade the correct answer.**

1.

$520 \div 100$

52 000	520	**5·2**

○ ○ ○

2.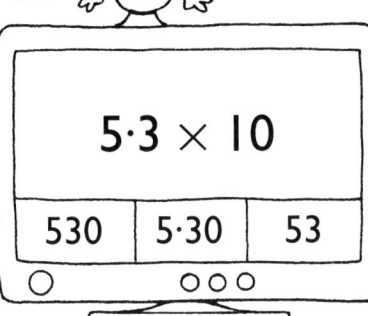

$5·3 \times 10$

530	5·30	53

○ ○ ○ ○

3.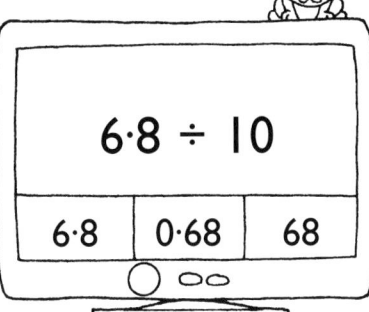

$6·8 \div 10$

6·8	0·68	68

○ ○ ○

4.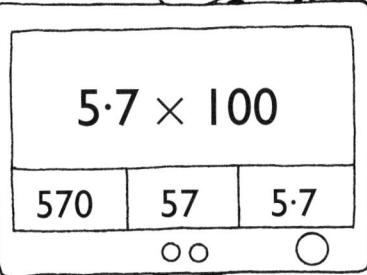

$5·7 \times 100$

570	57	5·7

○ ○ ○

5.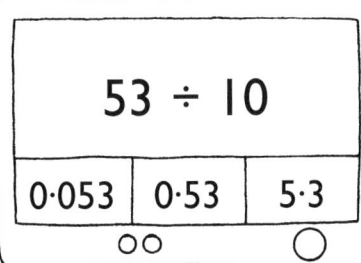

$53 \div 10$

0·053	0·53	5·3

○ ○ ○

6.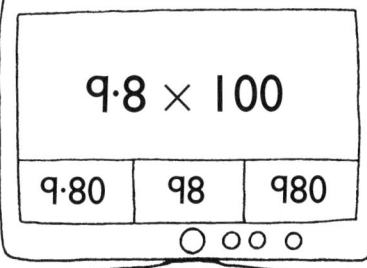

$9·8 \times 100$

9·80	98	980

○ ○ ○ ○

7.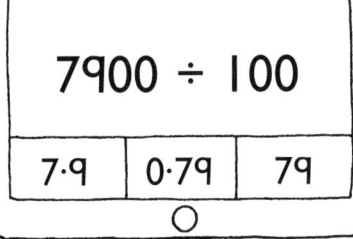

$7900 \div 100$

7·9	0·79	79

○

8.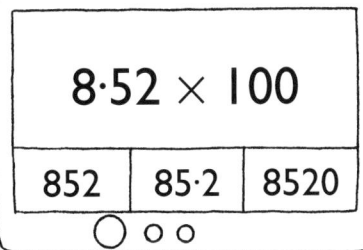

$8·52 \times 100$

852	85·2	8520

○ ○ ○

9.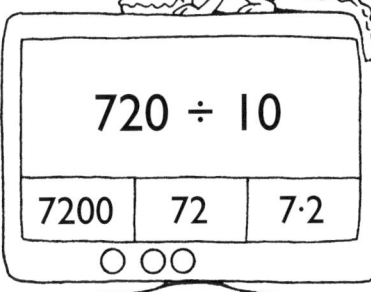

$720 \div 10$

7200	72	7·2

○ ○ ○

10.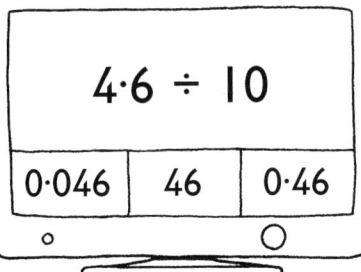

$4·6 \div 10$

0·046	46	0·46

○ ○

11.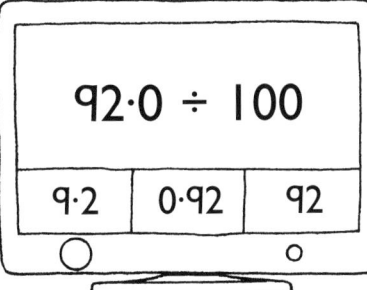

$92·0 \div 100$

9·2	0·92	92

○ ○

12.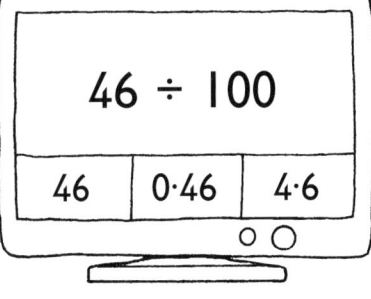

$46 \div 100$

46	0·46	4·6

○ ○

Now try this! • **Divide these numbers by** $\boxed{1000}$.

4500 = __4·5__ 6200 = _____ 8950 = _____ 11 320 = _____

6760 = _____ 3220 = _____ 30 050 = _____ 65 170 = _____

Teachers' note It is important that the children appreciate that it is the digits that move to the left or right when multiplying or dividing by 10, 100 or 1000, and that zeros are used as place holders to indicate the columns that are empty.

**Developing Numeracy
Mental Maths Year 6
© A & C BLACK**

Bungee jumping

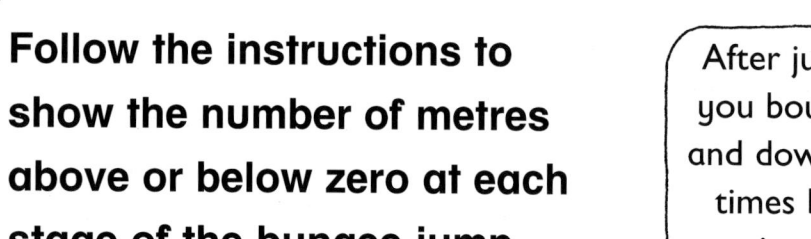

- **Follow the instructions to show the number of metres above or below zero at each stage of the bungee jump.**

After jumping, you bounce up and down a few times before coming to rest.

Jump 1
start at **12 m**

down 28 m	⁻16
up 24 m	8
down 11 m	⁻3
up 9 m	
down 8 m	
up 4 m	
down 3 m	
up 1 m	

Jump 2
start at **15 m**

down 30 m	
up 27 m	
down 21 m	
up 17 m	
down 13 m	
up 9 m	
down 5 m	
up 6 m	

Jump 3
start at **18 m**

down 31 m	
up 25 m	
down 14 m	
up 13 m	
down 10 m	
up 7 m	
down 5 m	
up 3 m	

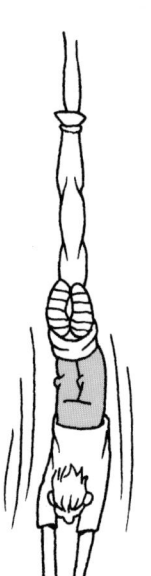

Jump 4
start at **22 m**

down 33 m	
up 31 m	
down 16 m	
up 14 m	
down 11 m	
up 7 m	
down 4 m	
up 1 m	

Now try this!

- **Find the difference between each pair of numbers.**

12 and ⁻5 _____	15 and ⁻8 _____	⁻4 and 17 _____
15 and ⁻7 _____	⁻9 and 13 _____	⁻11 and 14 _____
⁻6 and ⁻7 _____	⁻12 and ⁻5 _____	⁻4 and ⁻17 _____
⁻9 and ⁻5 _____	⁻16 and ⁻8 _____	⁻22 and ⁻10 _____

- **Make up two bungee jumping stories like those above.**

Teachers' note Draw a vertical number line and demonstrate how to move up and down to show rises and falls. Encourage the children to discuss strategies for finding the new heights without touching a number line, such as using zero as a stopover. When completing the sheet, less confident children may need a number line.

Developing Numeracy Mental Maths Year 6 © A & C BLACK

Square sequences

- **Write the number of spots in each square to make a sequence of** $\boxed{\text{square numbers}}$ **. Then find the difference between adjacent numbers.**

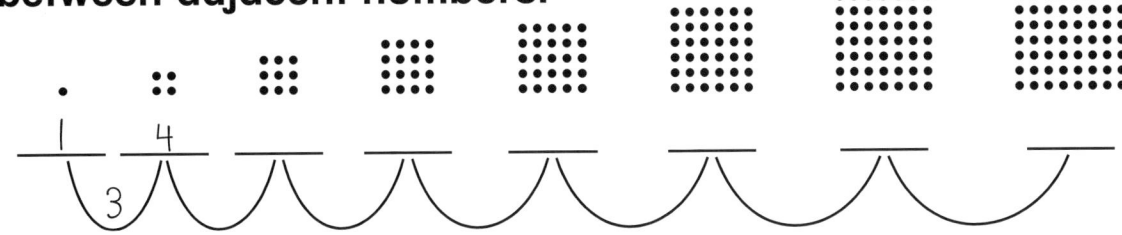

- **Use the differences to work out the next 12 square numbers.**

_____

- **Find the** $\boxed{\text{digital root}}$ **of each square number. To do this, keep adding the digits until you get a single-digit number.**

  Example: $49 \rightarrow 4 + 9 = 13 \rightarrow 1 + 3 = 4$ The digital root of 49 is **4**.

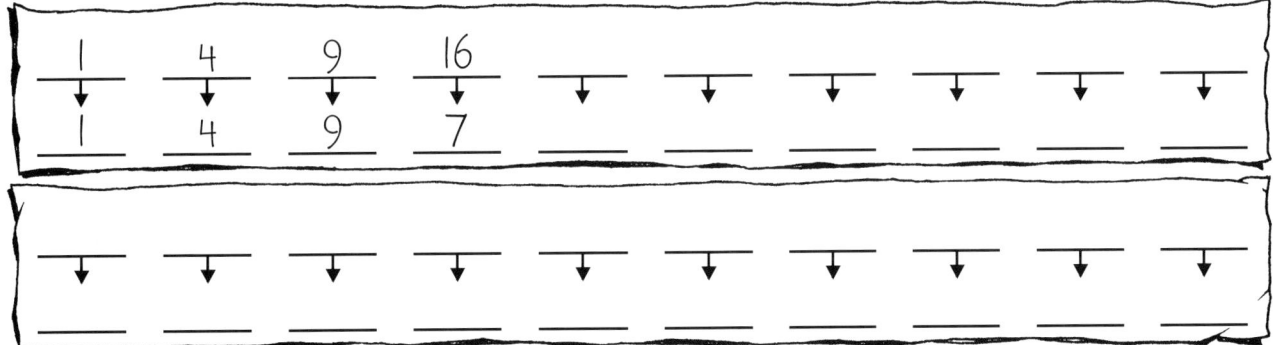

- **Describe what you notice.** _____

_____

- **Using what you have noticed, draw a ring around the numbers you think cannot be square numbers.**

  240   104 329   10 241   501 203

- **Continue this sequence of** $\boxed{\text{triangular numbers}}$ **.**

  1   3   6   10   15   ____   ____   ____   ____

- **Now find the** $\boxed{\text{sum}}$ **of adjacent numbers.**

  What do you notice?

Teachers' note Discuss that a square number is made by multiplying a number by itself, and revise the meaning of 'adjacent'. Demonstrate how to find the digital root of a number by adding its digits together until a single-digit number is reached: for example, the digital root of 48 is 3 (48 → 12 → 3). For the extension activity, ask the children to draw the triangles on a separate piece of paper.

**Developing Numeracy
Mental Maths Year 6
© A & C BLACK**

Slithering snakes

• **Complete the patterns.**

+0·5
0·5

+0·2
1·4

+0·5
1·2

+0·1
1   1·1

+0·25
3

+0·1
5·3

+0·1
6·2

−1
10·5

−0·1
8·45

• **Start from the number shown and count:**

on in 0·1s	9·63	_____	_____	_____	_____
on in 0·01s	4·66	_____	_____	_____	_____
back in 0·1s	10·35	_____	_____	_____	_____
back in 0·01s	9·02	_____	_____	_____	_____

Teachers' note At the start of the lesson, practise counting in steps of different sizes, such as 0·1, 0·2, 0·5 and 0·25, from any decimal with two decimal places.

Developing Numeracy Mental Maths Year 6 © A & C BLACK

Multiple shapes

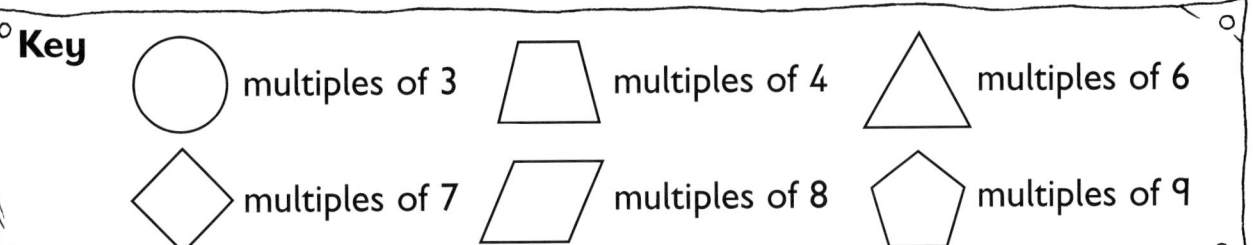

Key

◯ multiples of 3 ⬡(trapezoid) multiples of 4 △ multiples of 6

◇ multiples of 7 ▱ multiples of 8 ⬠ multiples of 9

- **Look at the key above. Write numbers in the shapes to make the correct answer.**

Example:

9 is a multiple of 3.

16 is a multiple of 8.

◯9 + ▱16 = 25

◯ + ◇ = 30   △ + ◇ = 45   ◯ + ⬡ = 24

▱ + △ = 68   ◯ + ⬠ = 42   ◇ + ⬡ = 44

◯ + ▱ = 59   ⬡ + ◇ = 56   ⬡ + ⬠ = 77

◯ + ◇ = 80   △ + ◇ = 77   ◯ + ⬡ = 87

▱ + △ = 100   ◯ + ⬠ = 108   ◇ + ⬡ = 105

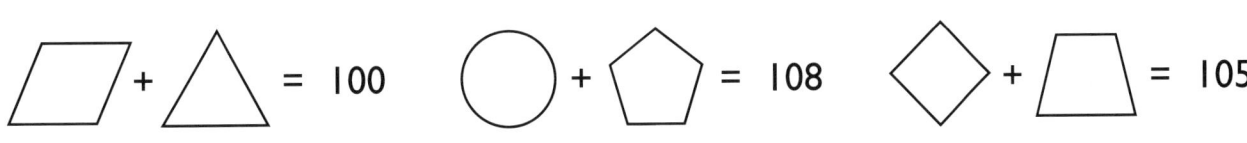

Some numbers are a multiple of more than one number.

- **Write the lowest number that completes each statement correctly.**

24 is a multiple of 8 and 3.

▱24 = ◯24   ⬠ = ⬡   ◇ = ◯

⬡ = ▱   ⬠ = △ = ⬡ = ◯

Teachers' note Revise multiples of 3, 4, 6, 7, 8 and 9, up to × 10 and beyond. Ensure the children understand that they can only write a number that is a multiple of the number shown in the key. Stress that the same shape can stand for different values in different questions. For the extension activity, some children may benefit from having a list of multiples of each number to refer to.

**Developing Numeracy
Mental Maths Year 6
© A & C BLACK**

All square

- **Write the** `square numbers` **in the boxes.**

1	4										
1×1	2×2	3×3	4×4	5×5	6×6	7×7	8×8	9×9	10×10	11×11	12×12

- **Write square numbers in these statements to make the correct answer. Make sure all your statements are different.**

$$\boxed{1} + \boxed{4} = 5 \qquad \boxed{} + \boxed{} = 8 \qquad \boxed{} + \boxed{} = 13$$

$$\boxed{} - \boxed{} = 5 \qquad \boxed{} - \boxed{} = 9 \qquad \boxed{} - \boxed{} = 13$$

$$\boxed{} + \boxed{} = 50 \qquad \boxed{} - \boxed{} = 15 \qquad \boxed{} - \boxed{} = 21$$

$$\boxed{} + \boxed{} = 50 \qquad \boxed{} - \boxed{} = 15 \qquad \boxed{} - \boxed{} = 21$$

Don't use the square number 1 in the questions below.

$$\boxed{} \times \boxed{} = 16 \qquad \boxed{} \times \boxed{} = 64 \qquad \boxed{} \times \boxed{} = 36$$

$$\boxed{} \times \boxed{} = 100 \qquad \boxed{} \times \boxed{} = 144 \qquad \boxed{} \times \boxed{} = 225$$

$$\boxed{} \div \boxed{} = 4 \qquad \boxed{} \div \boxed{} = 4 \qquad \boxed{} \div \boxed{} = 4$$

$$\boxed{} \div \boxed{} = 4 \qquad \boxed{} \div \boxed{} = 25 \qquad \boxed{} \div \boxed{} = 36$$

$$\boxed{} \div \boxed{} = 9 \qquad \boxed{} \div \boxed{} = 9 \qquad \boxed{} \div \boxed{} = 9$$

- **Find all the numbers between** $\boxed{1}$ **and** $\boxed{30}$ **that can be made by adding two square numbers.**

Teachers' note Check that the children understand what a square number is and ensure that they are beginning to know the sequence of square numbers to 12×12. Encourage them to try to explain why some patterns occur: for example, $4^2 \times 2^2$ is the same as $8^2 \times 1^2$, since $4^2 \times 2^2 = (2 \times 2)^2 \times 2^2 = (2 \times 2) \times (2 \times 2) \times 2 \times 2 = (2 \times 2 \times 2)^2 = 8^2$.

Developing Numeracy
Mental Maths Year 6
© A & C BLACK

Snookered

All the numbers on the balls are | prime factors | .

1. Multiply the numbers on the balls together. Write the answer on the cuff.

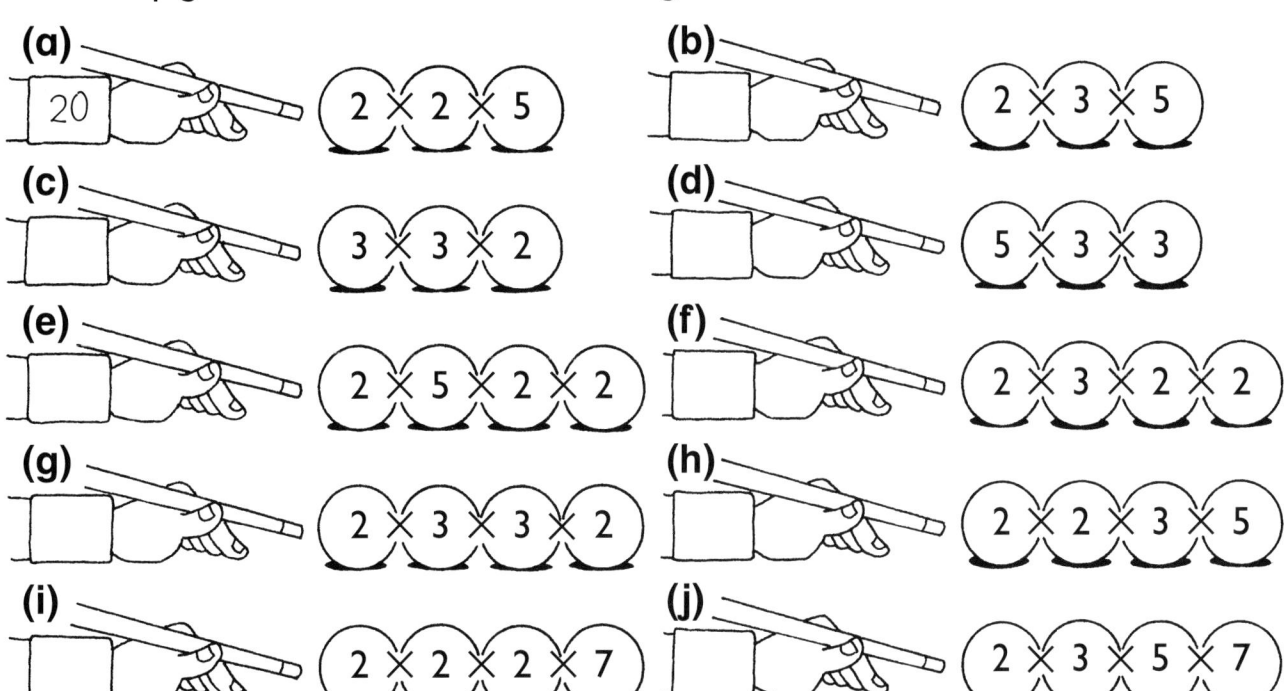

(a) 20   2 × 2 × 5

(b) ☐   2 × 3 × 5

(c) ☐   3 × 3 × 2

(d) ☐   5 × 3 × 3

(e) ☐   2 × 5 × 2 × 2

(f) ☐   2 × 3 × 2 × 2

(g) ☐   2 × 3 × 3 × 2

(h) ☐   2 × 2 × 3 × 5

(i) ☐   2 × 2 × 2 × 7

(j) ☐   2 × 3 × 5 × 7

2. Look at the number on the cuff. Write the correct prime factors

on the balls.

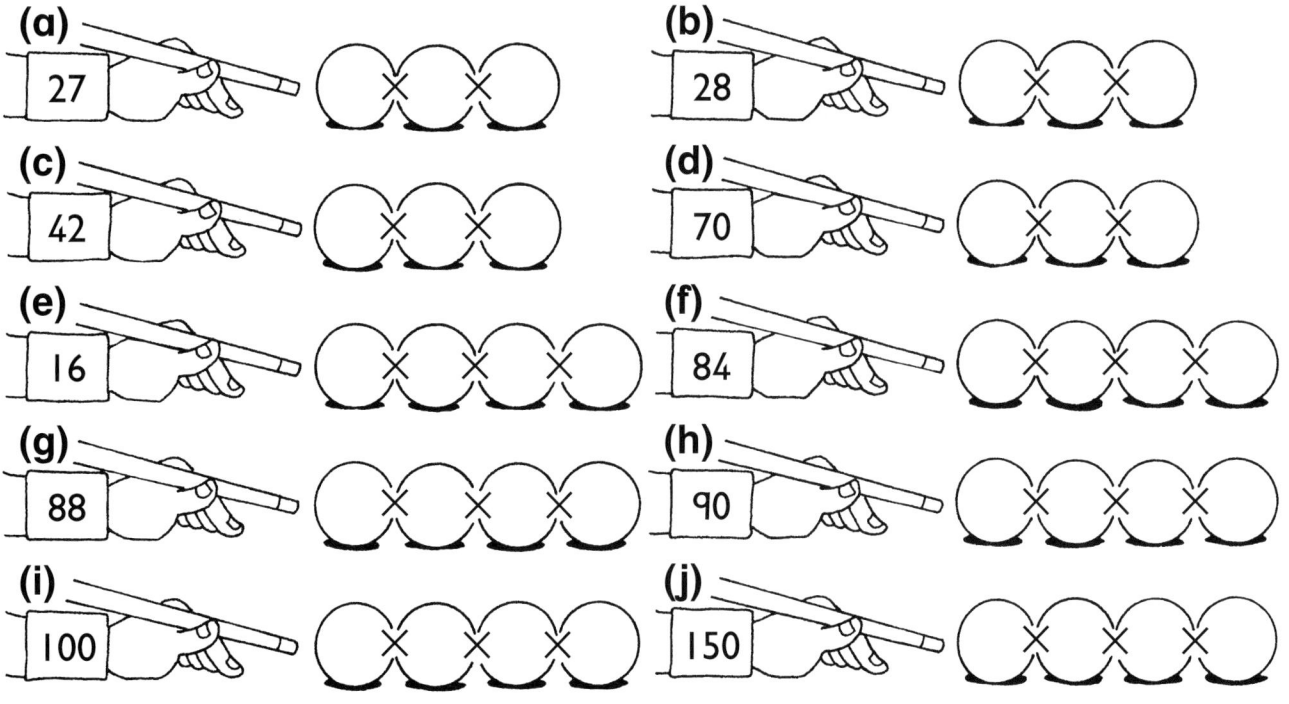

(a) 27   ◯ × ◯ × ◯

(b) 28   ◯ × ◯ × ◯

(c) 42   ◯ × ◯ × ◯

(d) 70   ◯ × ◯ × ◯

(e) 16   ◯ × ◯ × ◯ × ◯

(f) 84   ◯ × ◯ × ◯ × ◯

(g) 88   ◯ × ◯ × ◯ × ◯

(h) 90   ◯ × ◯ × ◯

(i) 100   ◯ × ◯ × ◯ × ◯

(j) 150   ◯ × ◯ × ◯

• **Choose other two-digit numbers that are not prime numbers. Find their prime factors in the same way.**

Teachers' note It is important that the children are already familiar with prime numbers before tackling this activity. Remind them that a prime number has only two factors: itself and one. List the prime numbers 2, 3, 5, 7, 11 on the board for the children to refer to. When doing question 2, the children can begin by finding one prime factor, then divide to find the others.

Developing Numeracy
Mental Maths Year 6
© A & C BLACK

Reveal the shape

- On each circle, join pairs of decimals with a total of $\boxed{10}$ or pairs of numbers with a total of $\boxed{100}$.
- Colour the shape inside the lines you have drawn. Write its name.

Use a ruler.

1.
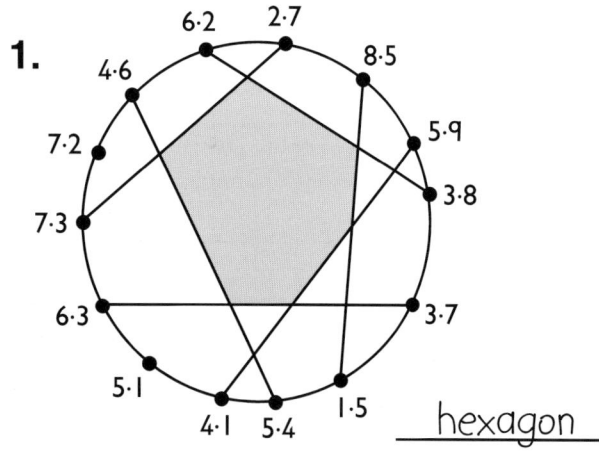
hexagon

2.
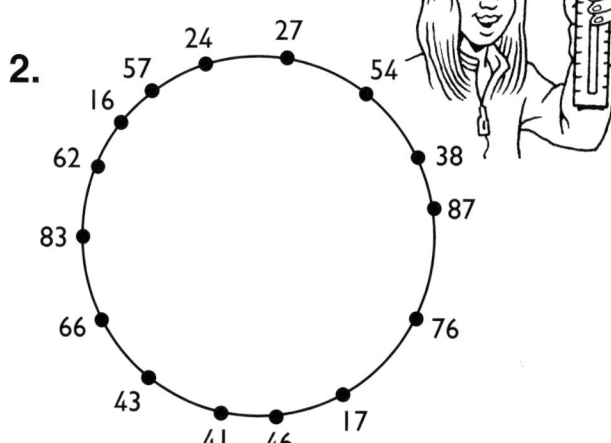
_____

3.
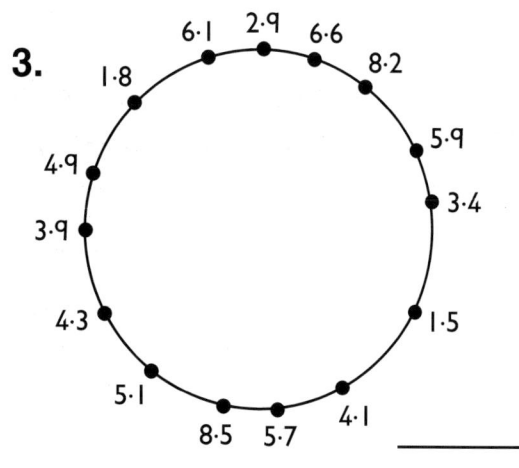
_____

4.
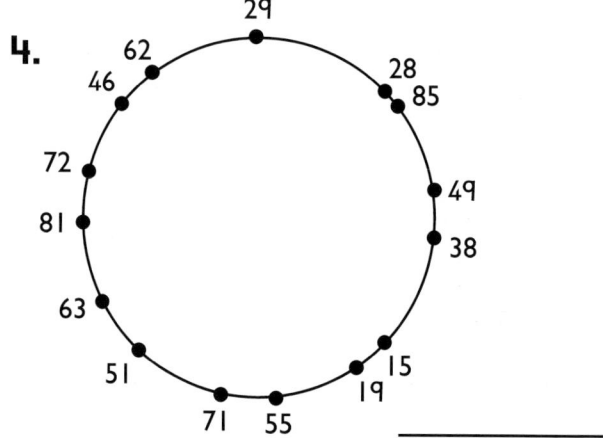
_____

5.
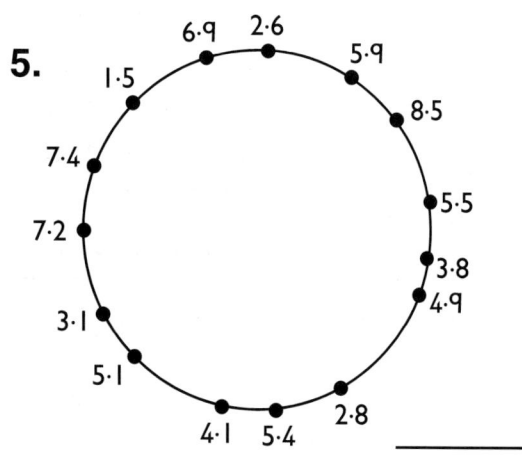
_____

6.
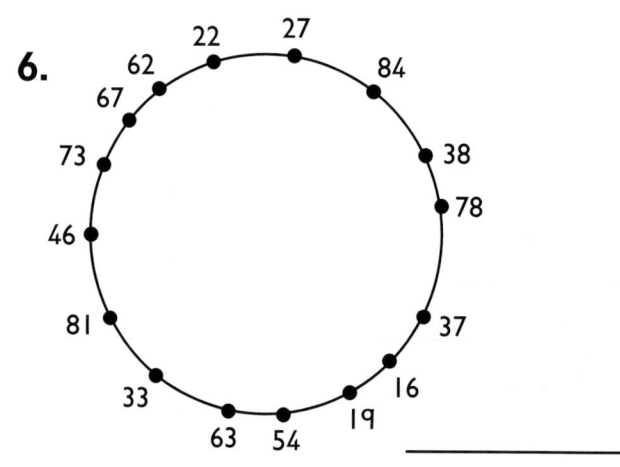
_____

Now try this!

- **Draw a large circle. Make your own 'hidden shape' puzzle for a partner to solve.**

Teachers' note Revise the shape names 'pentagon', 'hexagon', 'heptagon', 'octagon' and 'trapezium'. Remind the children that for pairs of numbers with a total of 100, the units digits will add to make 10, so the tens must total 90 only. Similarly, for decimals that add to make 10 the tenths digits will add to make 1 whole, so the units digits must total 9. Provide circle templates for the extension activity.

Developing Numeracy
Mental Maths Year 6
© A & C BLACK

Maths wizard

The maths wizard has a set of number cards. All the numbers are multiples of 10. He picks two cards which total ⟨ 1000 ⟩.

• Work out what numbers are on each pair of cards.

1. One number is 600 more than the other.

200 and 800

2. One number is 620 more than the other.

_____

3. One number is 660 more than the other.

_____

4. One number is 520 more than the other.

_____

5. One number is 180 more than the other.

_____

6. One number is 340 more than the other.

_____

7. One number is 820 more than the other.

_____

8. One number is 780 more than the other.

_____

q. One number is 220 more than the other.

_____

10. One number is 460 more than the other.

_____

• **Which pairs of numbers match these descriptions?**

The numbers total 1000. One is three times greater than the other.

The numbers total 1000. One is nine times greater than the other.

Now try this!

_____   _____

Teachers' note Do *not* provide a strategy for finding these solutions, but encourage the children to develop their own strategies (beginning, perhaps, with trial and error). Invite them to discuss their ideas with a partner and, towards the end of the lesson, encourage them to see that they can subtract the difference from 1000 and halve the answer to find the smaller number.

**Developing Numeracy
Mental Maths Year 6
© A & C BLACK**

Bugs in rugs

- **Find the bug's path across the rug by colouring <u>pairs</u> of adjacent numbers that have the difference shown.**

The numbers can be horizontally or vertically .

Start from the shaded number each time.

Count up from the smaller number to check your answer.

Difference of 8

4002	3994	5001	8001	3002
3997	5003	4993	3999	6006
8007	5997	6005	4002	5998
7999	6002	5996	8991	1996
6995	7003	9000	8992	2004

Difference of 10

5001	4998	3007	2997	6993
3991	5002	6004	6005	7003
6007	4002	5994	9004	8996
6997	3998	4008	8003	9006
1991	2001	5991	9002	8997

Difference of 12

8002	4002	6991	5999	7002
6992	7990	7003	6000	4007
7999	8002	3002	4002	3995
3006	2989	3001	7001	6996
5982	6000	5994	6006	7008

Difference of 15

7999	4002	3989	4004	7988
8002	7986	8001	5002	8003
2007	3010	2999	4999	3003
3005	2995	4996	5011	5010
6991	7001	5002	3987	4002

Difference of 21

3997	8020	5994	3971	4002
1987	7999	6006	5991	3981
2008	1994	8995	9015	8994
7985	8006	1988	2009	9005
8001	2995	3007	1999	2003

Difference of 25

3006	6995	7020	4991	5006
2975	3000	6991	5016	4999
3002	2998	7989	8012	7997
6980	7006	8004	7987	3007
7009	6984	8985	9010	2991

- **Make up your own bug trail, where adjacent numbers have a difference of** | 19 |.

Teachers' note Remind the children of the meaning of 'adjacent'. Demonstrate the use of an empty number line to find the difference between numbers either side of a multiple of 1000: for example, for 2005 – 1997, draw a line with 1997 at one end and 2005 at the other, with 2000 in between. Show that you can jump 3 to the multiple of 1000 and then 5 more, making the difference 8.

Developing Numeracy
Mental Maths Year 6
© A & C BLACK

Mind-reading tricks

- **Choose a four-digit number and follow the instructions. Record each step in the boxes. Try three different start numbers.**

What do you notice about the final numbers?

Choose your start number

Add 2999

Subtract 39

Add 3·1

Subtract 7·9

Subtract 899

Add 1·8

Subtract 58

Subtract your start number

4246		
7245		
7206		

- **Now follow these instructions.**

Choose your start number

Add 4999

Add 5·1

Subtract 99

Subtract 9·9

Subtract 699

Add 2·8

Subtract 199

Subtract your start number

Can you explain to a partner why the trick works?

Now try this!

- **Devise your own trick like this to perform on your friends. Make sure you know what the final number will be.**

Teachers' note At the start of the lesson, revise the strategy of adding or subtracting the nearest multiple of 10, 100 or 1000 and adjusting: for example, 1524 + 3999 can be found by adding 4000 and subtracting 1. Demonstrate that a similar strategy can be used for adding and subtracting decimals: for example, 28.4 − 2·9 can be found by subtracting 3 and adding 0·1.

**Developing Numeracy
Mental Maths Year 6
© A & C BLACK**

17

School fête

Choose a board. Score the target number to win a prize.

Sita and Ben have set up a darts stall.

Players choose a board and throw two or more darts to try to make a target number.

• Which target numbers up to 30 is it possible to score on each board?

Board 1
black ring scores 5
white ring scores 7

5 + 5 = 10
5 + 7 = 12
7 + 7 = 14
5 + 5 + 5 = 15
5 + 5 + 7 = 17
5 + 7 + 7 =

Board 2
black ring scores 4
white ring scores 5

4 + 4 =

Board 3
black ring scores 3
white ring scores 6

Board 4
black ring scores 6
white ring scores 2

Now try this!

• **Imagine you are running the stall. Which boards would you tell players to use if a prize is given for these target numbers? Discuss with a partner.**

(a) Target = 14   **(b)** Target = any even number

Teachers' note Provide extra paper for recording the totals. For the extension activity, talk about whether or not the children would wish someone to win a prize every time, sometimes or never. During the plenary, discuss which boards would produce the most interesting game at the school fête. Invite the children to explain the methods they used to answer the questions.

**Developing Numeracy
Mental Maths Year 6
© A & C BLACK**

Making a meal of it

- **Find the total of the numbers on each card.**
- **Cut out the cards. Sort them into piles with the same total. Then order the letters in each pile to spell a food or drink.**

70 70 80 60 **E** Total = _____	70 20 50 60 **I** Total = _____	40 90 80 30 **A** Total = _____
60 30 90 50 **E** Total = _____	80 50 40 70 **B** Total = _____	20 80 90 90 **K** Total = _____
10 50 90 80 **G** Total = _____	60 10 80 90 **S** Total = _____	90 80 70 60 **T** Total = _____
80 40 70 90 **A** Total = _____	30 60 70 40 **S** Total = _____	20 90 90 40 **E** Total = _____
60 10 90 70 **S** Total = _____	90 50 30 70 **N** Total = _____	60 80 60 80 **H** Total = _____
90 90 90 30 **T** Total = _____	70 90 70 70 **A** Total = _____	20 90 50 70 **G** Total = _____
10 50 80 60 **H** Total = _____	50 80 60 10 **C** Total = _____	80 50 80 70 **S** Total = _____
50 20 60 70 **P** Total = _____	80 50 80 90 **O** Total = _____	90 80 90 40 **S** Total = _____

Teachers' note Encourage checking by adding in a different order, and invite the children to say which order of adding they found easiest. As an extension activity, the children could create their own set of food cards by finding different sets of numbers with the same total and writing a letter on each card to spell a food. The cards can then be mixed with other sets and used for the same activity.

**Developing Numeracy
Mental Maths Year 6
© A & C BLACK**

Mean totals

• **Find the** | mean average | **of each set of cards, like this:**

☆ Find the total of the numbers by multiplying and adjusting.

☆ Then divide the total by the number of cards to find the **mean**.

1. | 71 | 70 | 69 | 74 | $4 \times 70 = 280$

+1   −1 +4

Total = 284   Mean = 284 ÷ 4 = 71

Each of these four numbers is about **70**. So, I can multiply by 70 and then adjust.

2. | 49 | 52 | 54 | 49 |

Total =   Mean =

3. | 22 | 24 | 23 | 25 | 21 |

Total =   Mean =

4. | 64 | 62 | 59 | 63 |

Total =   Mean =

5. | 62 | 60 | 57 | 59 | 57 |

Total =   Mean =

6. | 15 | 18 | 13 | 14 |

Total =   Mean =

7. | 21 | 19 | 21 | 16 | 18 |

Total =   Mean =

8. | 89 | 80 | 88 | 82 | 81 |

Total =   Mean =

9. | 70 | 68 | 73 | 74 | 69 | 72 |

Total =   Mean =

10. | 77 | 76 | 78 | 81 | 77 | 79 |

Total =   Mean =

11. | 93 | 94 | 96 | 101 | 92 | 94 |

Total =   Mean =

Now try this!

• **These ten numbers have a mean of** | 92 |. **Fill in the missing number.** | 91 | 86 | 94 | 92 | 91 | 94 | 99 | 89 | 83 | |

Teachers' note Demonstrate how the totals can be found by multiplying and then adjusting. Discuss that '+ 1' and '− 1' cancel each other out. Ensure the children realise that they can round up or down when choosing the number for the approximation.

Developing Numeracy Mental Maths Year 6 © A & C BLACK

20

Groovy grids

• **Complete the grid. Then write the sequence of numbers in the circles. Describe the sequence.**

Add or subtract as you move along the rows and up or down.

1.

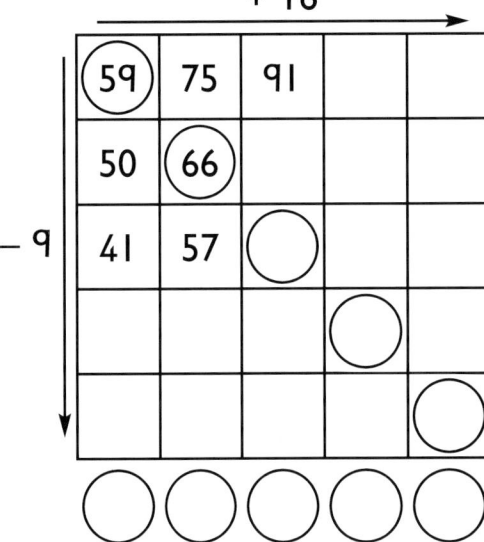

2.

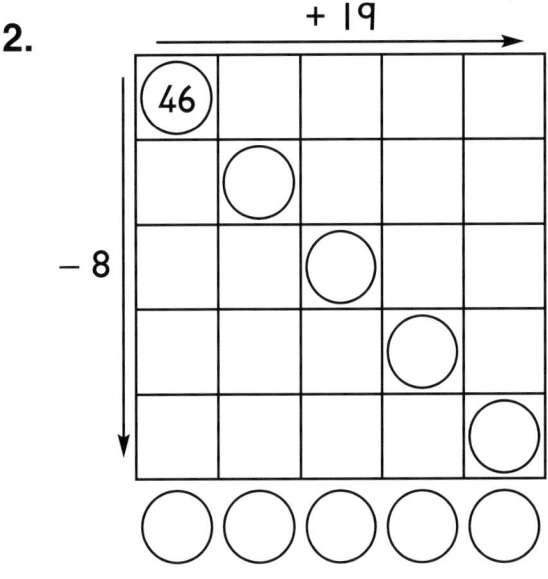

_____

_____

3.

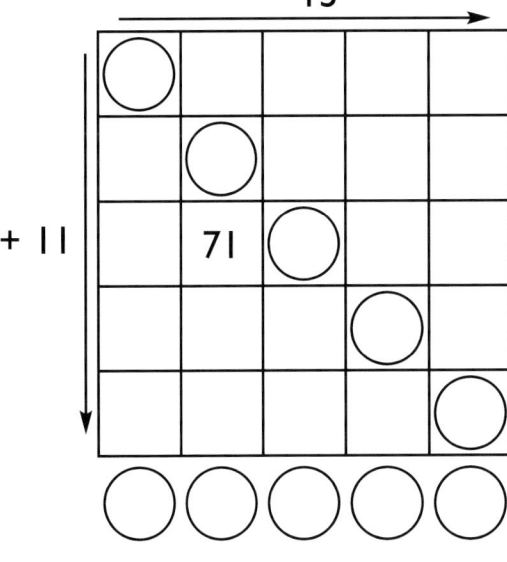

4.

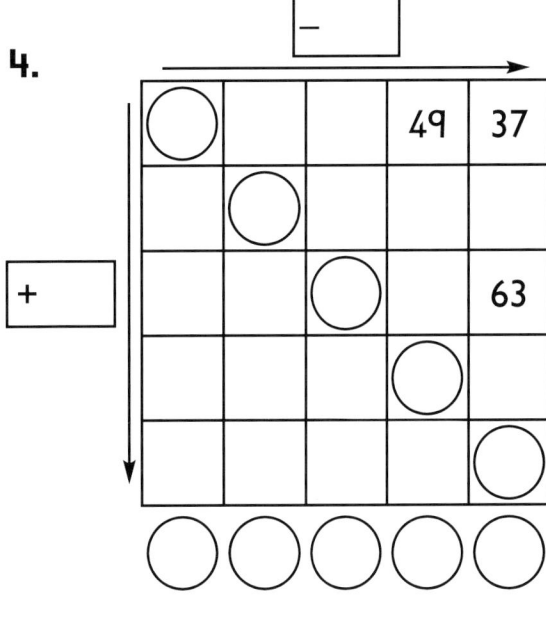

_____

_____

• **Draw your own grid with** (57) **in the centre. Choose rules for the rows and columns and complete the grid.**

Teachers' note Ensure the children understand how to complete the grids. Encourage discussion of ways to predict and find answers, using inverse operations where necessary. Discuss the importance of checking mental calculations, working both across and down as a check. Ask the children to discuss the relationship between the diagonal numbers and the column and row rules.

Developing Numeracy Mental Maths Year 6 © A & C BLACK

Hamster wheel

- ## You need two counters and a dice.

☆ Place two counters in different positions on the wheel.

☆ Roll a dice and move **one** of your counters in a clockwise direction.

☆ Find the total of the two numbers that your counters are on.
   If it is in the grid below, cross it off.

☆ Keep going until you have crossed off five totals in a line.

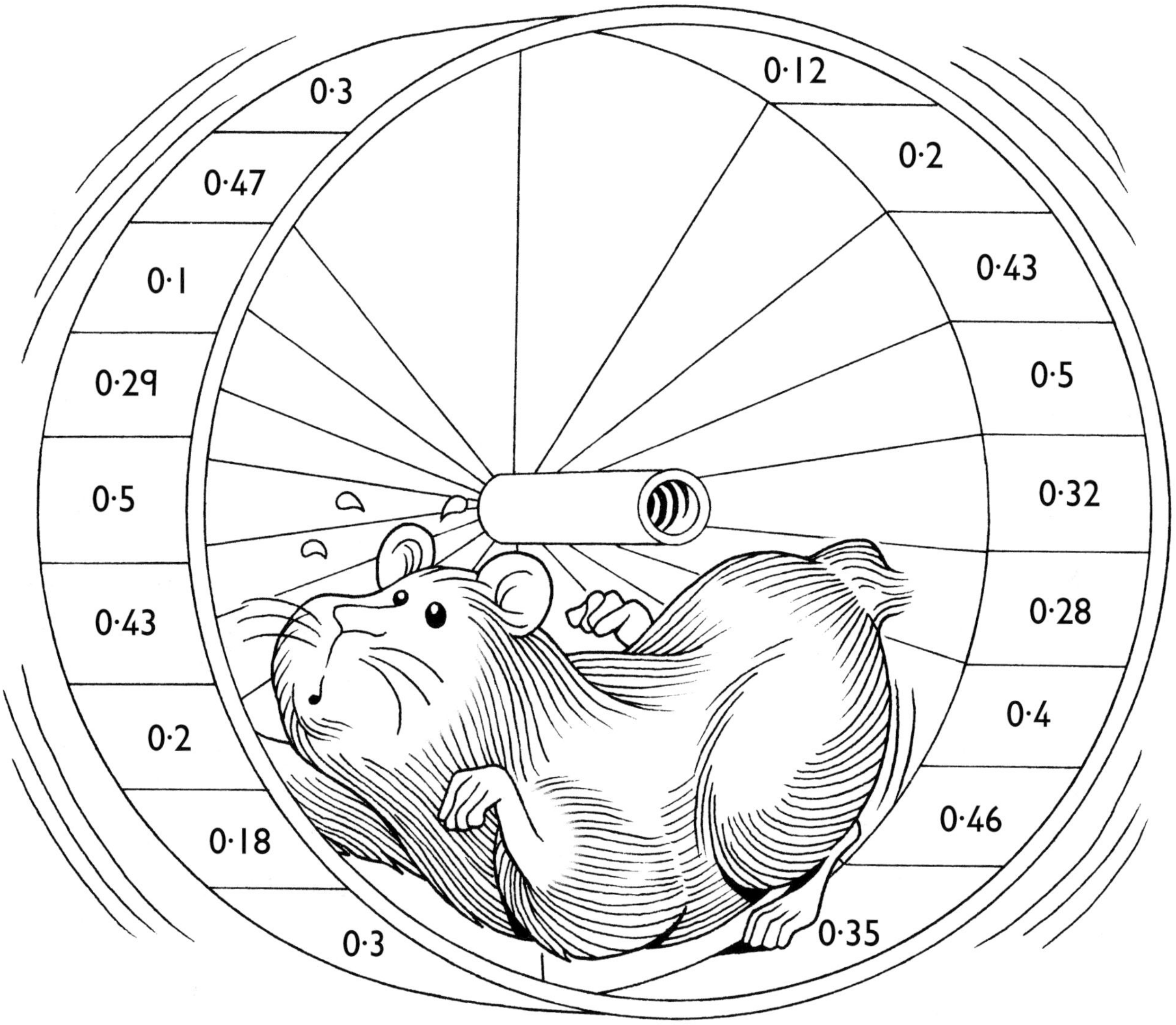

0·3   0·12   0·2   0·47   0·43   0·1   0·5   0·29   0·32   0·5   0·28   0·43   0·4   0·2   0·46   0·18   0·35   0·3

0·6	0·97	0·79	0·4	0·39	0·57	0·44	0·55
0·75	0·28	0·6	0·72	0·46	0·13	0·58	0·5
0·68	0·49	0·65	0·4	0·79	0·69	0·62	0·57
0·53	0·76	0·47	0·48	0·97	0·91	0·42	0·62
0·4	0·32	0·22	0·85	0·86	0·93	0·61	0·78
0·9	0·7	0·82	0·79	0·67	0·81	0·56	0·71

Teachers' note The children should record the questions and answers on a separate piece of paper. As an extension, ask the children to draw their own grid and write in *differences* between pairs of numbers in the wheel. This grid can be used to play the game again, this time finding the difference between the numbers covered by the counters. The children could play this in pairs.

**Developing Numeracy
Mental Maths Year 6
© A & C BLACK**

Space landing

- **Look at the decimal on the spacecraft.**
- **Write what you must add and subtract to reach the two nearest whole numbers.**

1. 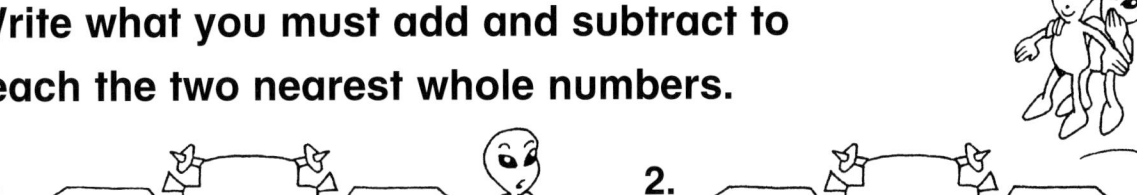 -0.3 | 4·3 | $+0.7$ | 4 | 5

2. 8·6 | 8 | 9

3.  7·2

4. 5·4

5.  3·9

6. 3·1

7.  6·39

8. 2·63

9. 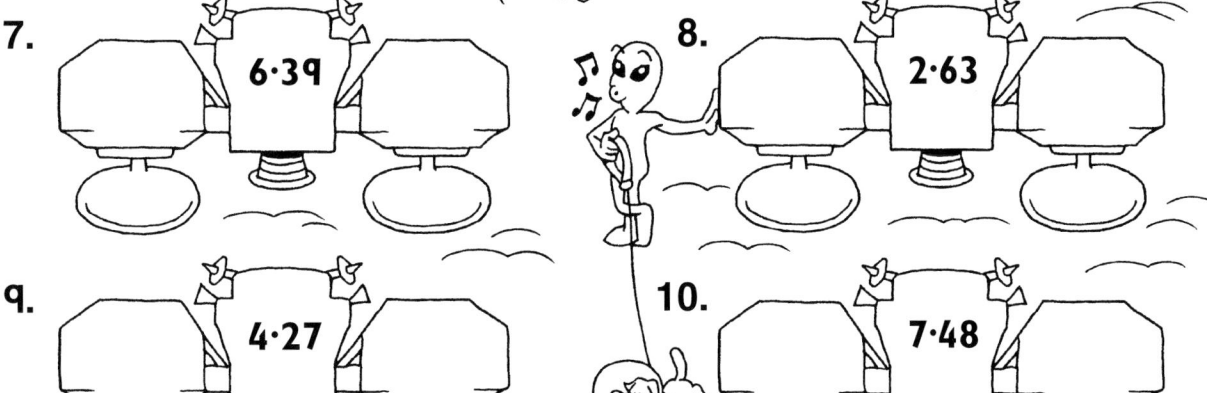 4·27

10.  7·48

The decimal on this spacecraft is a number between 1 and 10.
- **Write all the possible decimals it could be.**

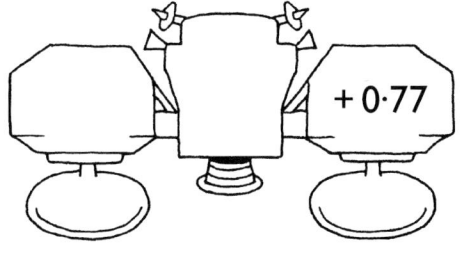 $+0.77$

Teachers' note Encourage the children to use their knowledge of number pairs with a total of 10 or 100. Watch out for children making mistakes such as 4·27 + 0·83 = 5.

Developing Numeracy
Mental Maths Year 6
© A & C BLACK

Magic squares

- **Look at this magic square. Multiply each number by the numbers below to make new magic squares.**

6	7	2
1	5	9
8	3	4

> In a magic square, all the rows, columns and diagonals have the same total. Check yours.

× 4

24		

× 8

× 6

× 7

× 3

× 9

Now try this!

- **For this magic square, divide each number by 2 and add 1. Is the new grid a magic square?**

60	70	20
10	50	90
80	30	40

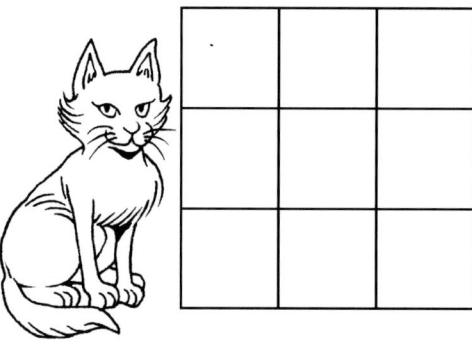

- **Now try these rules:** ÷ 10 + 10   ÷ 5 × 3   ÷ 10 − 5

  Do they all make magic squares?

- **Try some other rules of your own. What do you find?**

Teachers' note The main activity on this sheet provides practice of multiplication facts up to 10 × 10. At the start of the lesson, ask the children to arrange the digits 1 to 9 in a 3 × 3 grid so that each row, column and diagonal totals 15.

Developing Numeracy Mental Maths Year 6 © A & C BLACK

The X-factor

The two lines that cross at the central circle show pairs of numbers with the same ⬚ product .
Other products are shown in the outer circles.

• Complete each pattern.

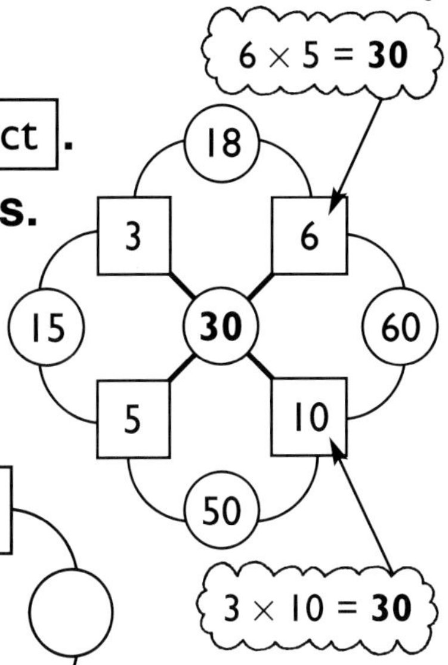

$6 \times 5 = 30$

18
3   6
15   **30**   60
5   10
50

$3 \times 10 = 30$

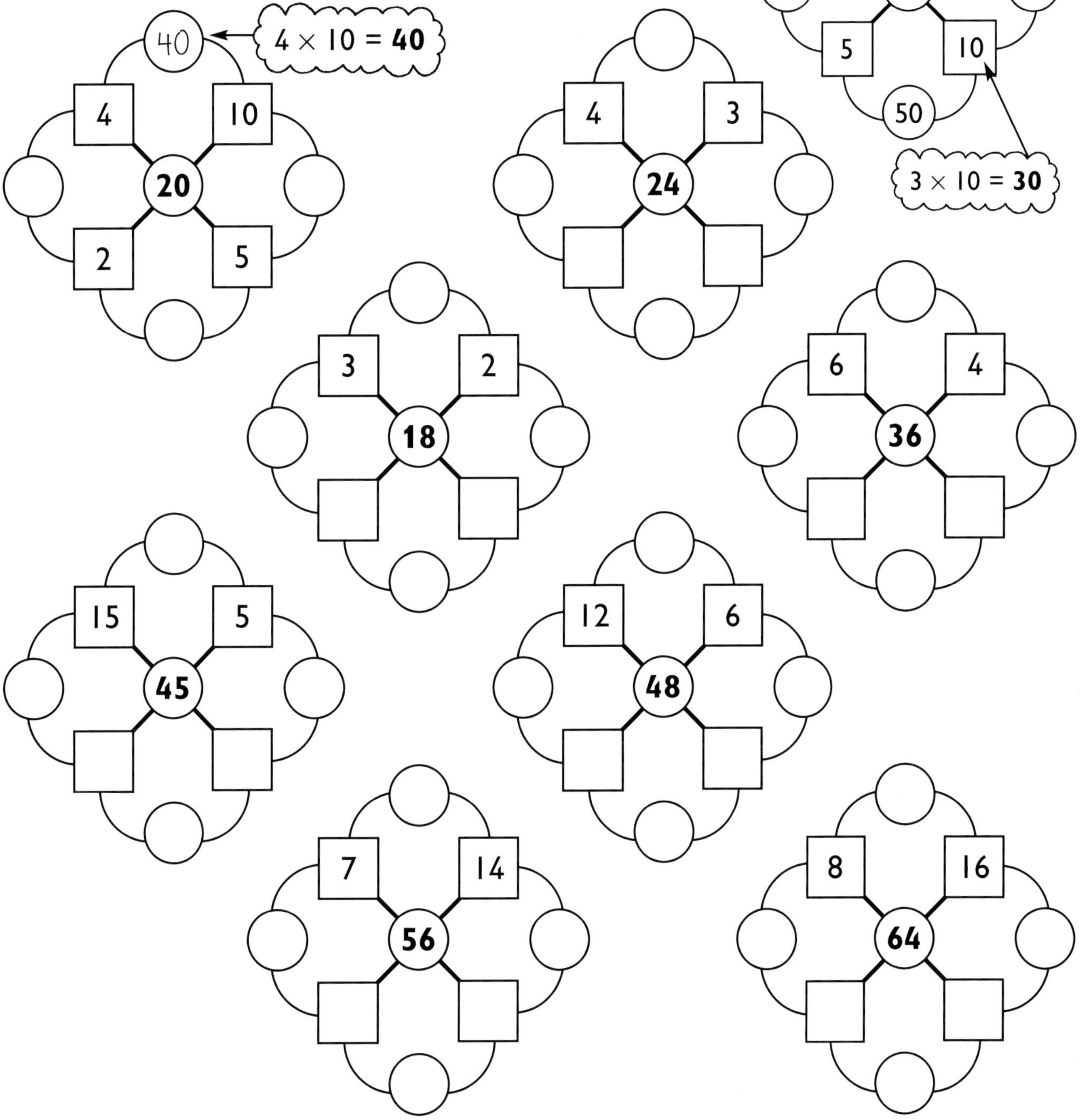

40    $4 \times 10 = 40$

4   10
20
2   5

4   3
24

3   2
18

6   4
36

15   5
45

12   6
48

7   14
56

8   16
64

• Draw five more patterns using these centre numbers.

(50) (100) (84) (90) (96)

Teachers' note Revise that 'product' means that numbers are multiplied together. As a further activity, the children could use a calculator to explore the relationship between the outer products and the central product. This can be done by multiplying all four outer products together and pressing the square root key twice. Ask the children what they notice.

**Developing Numeracy
Mental Maths Year 6
© A & C BLACK**

Now try this!

Speed tests

• **Time yourself as you complete each test. Do you get quicker?**

Test 1

1. Multiply nine by six ☐
2. Four lots of seven ☐
3. Sixty squared ☐
4. Eight times nine ☐
5. Five fives ☐
6. Share forty-two between six ☐
7. Six groups of eight ☐
8. Twenty-eight divided by four ☐
9. How many fours in thirty-six? ☐

Time _____

Test 2

1. Multiply six by eight ☐
2. Six lots of nine ☐
3. Forty squared ☐
4. Twenty twenties ☐
5. Seven sevens ☐
6. Share forty-eight between four ☐
7. Five groups of nine ☐
8. Thirty-five divided by five ☐
9. How many sixes in thirty-six? ☐

Time _____

Test 3

1. Multiply six by seven ☐
2. Eight lots of six ☐
3. Fifty squared ☐
4. Seven times nine ☐
5. Six sixes ☐
6. Share fifty-four between six ☐
7. Seven groups of ten ☐
8. Forty-two divided by seven ☐
9. How many eights in seventy-two? ☐

Time _____

Test 4

1. Multiply nine by seven ☐
2. Six lots of seven ☐
3. Ninety squared ☐
4. Seven times four ☐
5. Eight eights ☐
6. Share sixty-three between nine ☐
7. Nine groups of eight ☐
8. Fifty-four divided by nine ☐
9. How many fives in forty-five? ☐

Time _____

Teachers' note At the start of the lesson, revise that squaring a number means multiplying it by itself. The children could use stopwatches or clocks to time themselves. As an extension activity, ask the children to make up their own round of questions, using the 6, 7 and 8 times tables, for a partner to answer.

**Developing Numeracy
Mental Maths Year 6**
© **A & C BLACK**

Sweet 16

- **Investigate these puzzles.**

Be systematic so that you find all the combinations.

1. **Two** positive integers have a total of **16**.

   What could their product be?

   $1 \times 15 = 15$ $2 \times 14 =$ $3 \times 13 =$

   Tick the pair of numbers which has the highest product.

2. **Three** positive integers have a total of **16**.

   What could their product be?

   $1 \times 1 \times 14 = 14$ $1 \times 2 \times 13 =$ $1 \times 3 \times 12 =$

   Tick the set of numbers which has the highest product.

3. **Four** positive integers have a total of **16**.

   What could their product be?

   $1 \times 1 \times 1 \times 13 = 13$

Record your answers on a separate piece of paper.

   Tick the set of numbers which has the highest product.

Now try this!

- **Do the same for numbers with a total of** $\boxed{18}$ **.**
  Predict which set of numbers will have the highest product each time.
- **Try including decimals such as 10·5 and 7·5.**

Teachers' note This investigation provides practice of mental multiplication and can be adapted for any ability by altering the total number. Remind the children that 'product' means that numbers are multiplied together. Revise that an integer is a positive or negative whole number.

**Developing Numeracy
Mental Maths Year 6
© A & C BLACK**

The *a* and *b* game

Play this game with a partner.

☆ Cut out the cards. Spread them out face down.

☆ Each player chooses a small card and turns it over.

☆ Turn over **one** long card to show the value of *a* and *b*. Work out how much your card is worth. The player with the highest value keeps the two small cards. Turn the long card face down again.

☆ The winner is the player who collects the most small cards.

a = 5	*b* = 6	*a* = 6	*b* = 7
a = 4	*b* = 4	*a* = 5	*b* = 8
a = 9	*b* = 2	*a* = 2	*b* = 9
a = 6	*b* = 3	*a* = 7	*b* = 4
a = 8	*b* = 1	*a* = 6	*b* = 5

$a \times b$	$4a$	$a + b$	$6b$	$8a$
$a \times b \times 2$	$2a + b$	$5a - b$	$3b$	$7a$
$a + 5b$	$3a + b$	$9a - b$	$a \times b \div 2$	$5b$
$6a$	$a + 3b$	$30 - a$	$45 - b$	$8b$
$2a + 2$	$10a + 5$	$8b - a$	$(a + b) \times 2$	$(a + b) \times 3$
$2a + 2b$	$5a + b$	$9b$	$b \times a$	$7a - 10$

Teachers' note This game provides practice of multiplication facts up to 10 × 10. Ensure the children understand that the letters *a* and *b* stand for numbers, and explain that the notation 2*a* means '2 times *a*' or '2 lots of *a*'. Some cards could be removed or altered to make the game suitable for all abilities.

**Developing Numeracy
Mental Maths Year 6
© A & C BLACK**

Double crosser

- **Use the clues to complete the crossnumber puzzle.**

Look carefully for addition or subtraction signs in the clues.

Across

1 double 50 + double 3

3 double 64

6 double 150 + double 52

7 double 275 − double 25

8 half of 136 − double 29

9 half of 150 + double 11

11 half of 128

13 double 72 + double 72

15 half of 98 − double 17

16 half of 96

17 half of 240 − half of 160

18 double 18 − half of 44

19 half of 182 − half of 142

20 double 87 − half of 142

21 double 77 − double 59

22 half of 132 − half of 106

23 half of 76 − double 13

25 double 370 − half of 480

26 half of 220 − double 10

27 half of 94 − half of 26

28 half of 198

Down

1 half of 32

3 double 60 + half of 58

5 double 56 + half of 176

10 double 50 − half of 56

12 double 36 + double 36

18 half of 42 + double 55

20 double 85 − half of 2

24 half of 46

2 double 320

4 half of 170

8 half of 280 + half of 10

11 half of 122

14 half of 168

19 double 150 − half of 70

22 double 78 − half of 94

The crossnumber grid: cell **1** contains `1 0 6`.

- **Now write different doubling and halving clues for the answers in the crossnumber puzzle.**

Example: **Across**

1 double 100 − half of 188

Teachers' note At the start of the lesson, revise doubling and halving strategies, including using partitioning: for example, half of 178 = half of 170 + half of 8 = 85 + 4 = 89. As the children work through the crossnumber puzzle, encourage them to look for patterns such as double 77 − double 59 is the same as double (77 − 59) = double 18 = 36.

Developing Numeracy
Mental Maths Year 6
© A & C BLACK

Celestial symbols

The letters *a* and *b* always stand for multiples of 100.

• Use the key to find the value of each symbol.

1. *a* = 1800 and *b* = 2200

   ☆   ☀   ☽

__8000__   _____   _____   _____

2. *a* = 3600 and *b* = 2800

      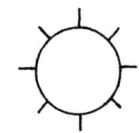   

_____   _____   _____   _____

3. *a* = 4400 and *b* = 6200

      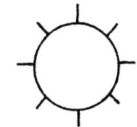   

_____   _____   _____   _____

4. *a* = 7800 and *b* = 2500

      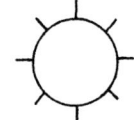   

_____   _____   _____   _____

• Find the value of *a* and *b*, if the four symbols are:

   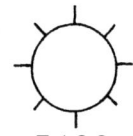   ☀

8000   2000   5600   4400

a = _____

b = _____

Teachers' note Ensure the children understand that the letters *a* and *b* stand for numbers.
Encourage the children to use known facts to derive facts they do not know by heart. Ask them to
describe their strategies to a partner: for example, 'I know that double 36 is 72, so double 3600 must
be 7200.'

**Developing Numeracy
Mental Maths Year 6
© A & C BLACK**

Decisions, decisions

- To answer these multiplications, double one number and halve the other. Think carefully about which to double and which to halve.

1.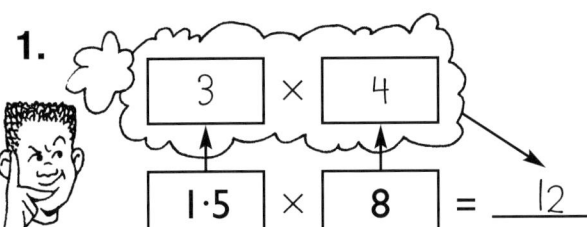

| 3 | × | 4 |

| 1·5 | × | 8 | = 12

2.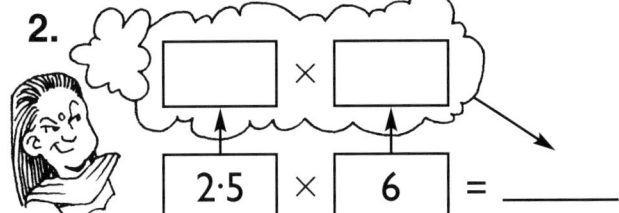

| | × | |

| 2·5 | × | 6 | = _____

3.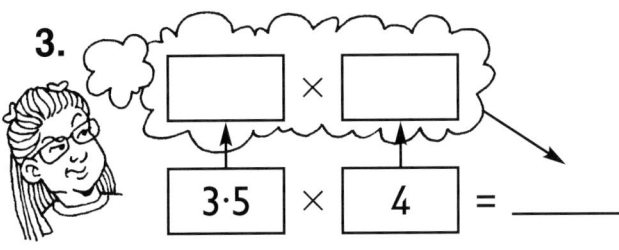

| | × | |

| 3·5 | × | 4 | = _____

4.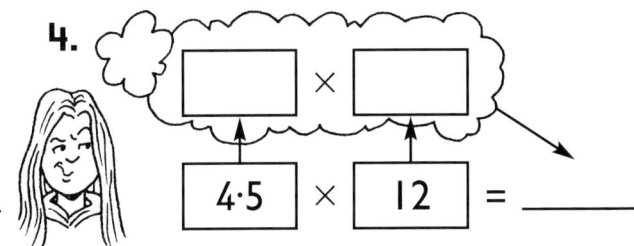

| | × | |

| 4·5 | × | 12 | = _____

5.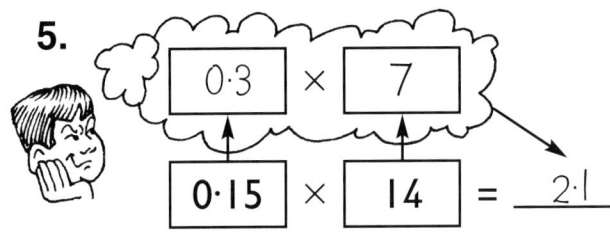

| 0·3 | × | 7 |

| 0·15 | × | 14 | = 2·1

6.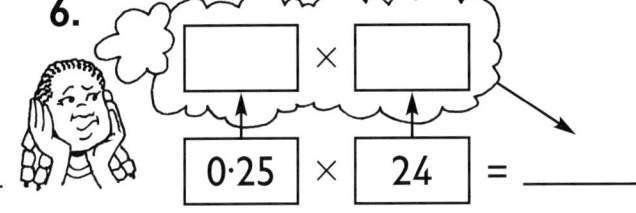

| | × | |

| 0·25 | × | 24 | = _____

7.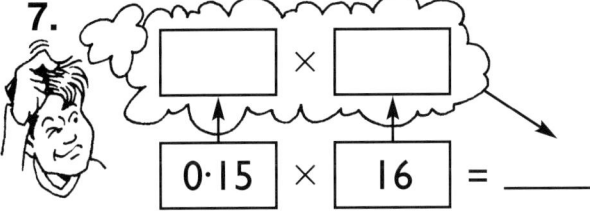

| | × | |

| 0·15 | × | 16 | = _____

8.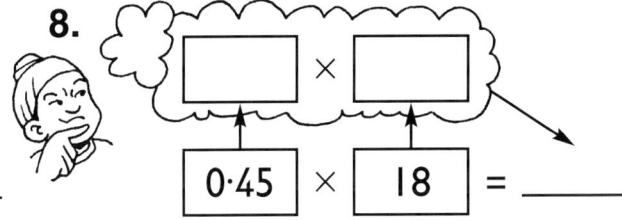

| | × | |

| 0·45 | × | 18 | = _____

9.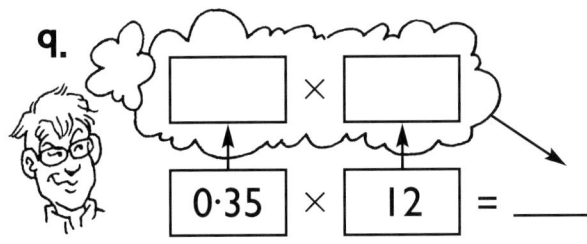

| | × | |

| 0·35 | × | 12 | = _____

10.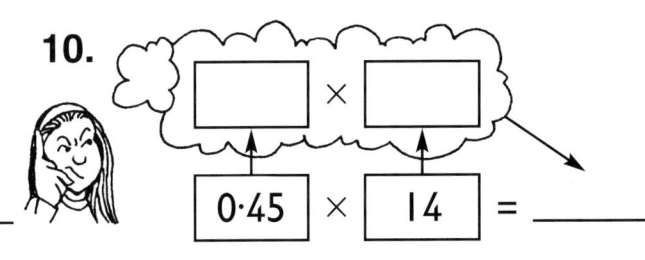

| | × | |

| 0·45 | × | 14 | = _____

Now try this!

- **Solve these multiplications in the same way.**

  3·5 × 14 = _____   4·5 × 16 = _____   5·5 × 18 = _____

  5·5 × 1·8 = _____   1·5 × 1·4 = _____   4·5 × 0·8 = _____

  0·35 × 16 = _____   0·15 × 12 = _____   0·25 × 14 = _____

  0·55 × 14 = _____   0·45 × 12 = _____   0·45 × 18 = _____

Teachers' note At the start of the lesson, revise multiplying decimals by single-digit numbers using tables facts as a starting point (for example, 0·3 × 9, 0·6 × 5 and 0·8 × 7 using knowledge of 3 × 9, 6 × 5 and 8 × 7). Demonstrate how to double one number and halve the other, and show that this produces the same answer as the original question.

Developing Numeracy Mental Maths Year 6 © A & C BLACK

Related times

- ## Work with a partner.

☆ Cut out the cards. Spread them out face up.

☆ Take turns to choose a card and find a related calculation. Explain to your partner how you know the cards are related.

☆ Keep going until all the cards are in pairs.

Look at the two grey cards.
4×35 is the same as $4 \times 5 \times 7$.

4×35	6×14	7×18
9×16	$8 \times 7 \times 2$	6×24
$7 \times 9 \times 2$	9×18	8×12
$9 \times 8 \times 2$	$8 \times 6 \times 2$	$6 \times 2 \times 12$
$4 \times 5 \times 7$	$9 \times 3 \times 6$	$6 \times 2 \times 7$
7×16	$9 \times 7 \times 2$	8×14
5×24	$5 \times 2 \times 36$	$7 \times 8 \times 2$
9×14	15×14	$3 \times 5 \times 2 \times 7$
5×72	35×18	$5 \times 2 \times 12$
18×15	16×55	45×12
$45 \times 2 \times 6$	$7 \times 2 \times 5 \times 9$	14×45
$35 \times 2 \times 9$	$8 \times 2 \times 5 \times 11$	$9 \times 2 \times 5 \times 3$

Teachers' note Remind the children that multiplication can be done in any order: for example, $9 \times 2 \times 5 \times 3$ has the same answer as $2 \times 5 \times 9 \times 3 = 10 \times 27 = 270$. As an extension, ask the children to calculate the answer to each pair, using either card.

**Developing Numeracy
Mental Maths Year 6
© A & C BLACK**

Aerobatics

To multiply by $\boxed{49}$ or $\boxed{51}$, multiply by $\boxed{50}$ first and adjust.

• **Use this method to complete the questions.**

1.

13 × 49 =
× 50 = 650
× 51 =

To multiply by 50 you can multiply by 100 and halve it.

2.
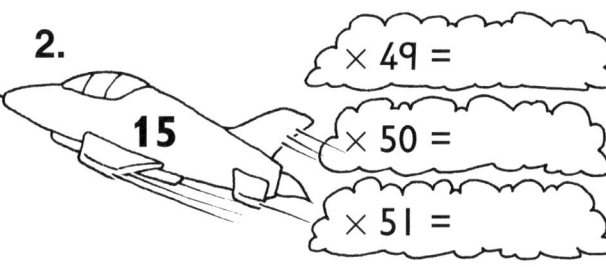

15 × 49 =
× 50 =
× 51 =

3.

16 × 49 =
× 50 =
× 51 =

4.
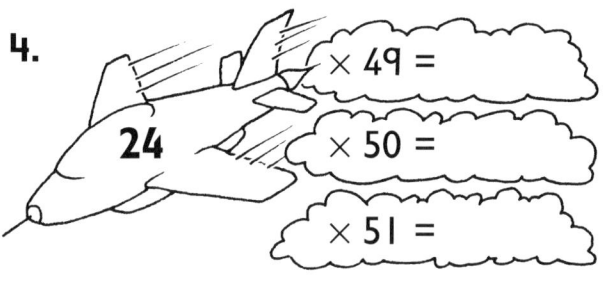

24 × 49 =
× 50 =
× 51 =

5.
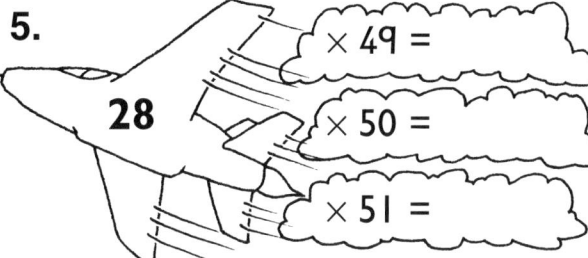

28 × 49 =
× 50 =
× 51 =

6.
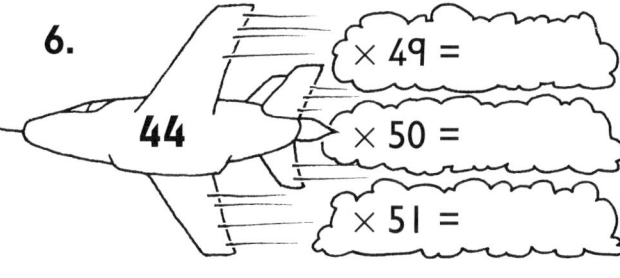

44 × 49 =
× 50 =
× 51 =

7.

26 × 49 =
× 50 =
× 51 =

8.
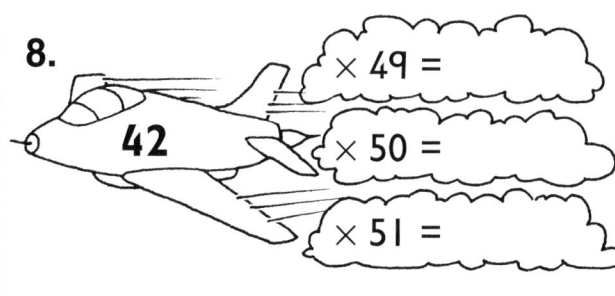

42 × 49 =
× 50 =
× 51 =

9.
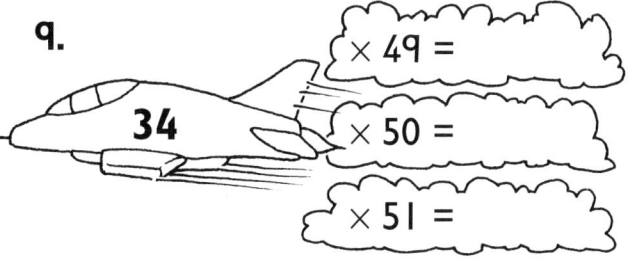

34 × 49 =
× 50 =
× 51 =

Now try this!

• **Use a similar method to answer these multiplications.**

$14 \times 49 =$ _____	$22 \times 51 =$ _____	$18 \times 49 =$ _____
$25 \times 51 =$ _____	$36 \times 49 =$ _____	$13 \times 99 =$ _____
$18 \times 99 =$ _____	$22 \times 99 =$ _____	$32 \times 101 =$ _____

Teachers' note Demonstrate this strategy at the start of the lesson. Show that to multiply a number by 49 we can multiply by 50 and then subtract the number. To multiply a number by 51 we can multiply by 50 and then add the number. The children can check their answers by making sure that the three answers in each set go up in steps of equal size (the size of the number on the plane).

Developing Numeracy
Mental Maths Year 6
© A & C BLACK

33

Time for tables

• **Complete these tables.**

Use the first two columns to help you complete the third column.

	× 10	× 7	× 17
1	10	7	17
2	20	14	34
3	30		
4	40		
5			
6			
7			
8			
9			
10			

	× 6	× 60	× 66
1	6	60	66
2	12	120	132
3			
4			
5			
6			
7			
8			
9			
10			

	× 20	× 8	× 28
1			
2			
3			
4			
5			
6			
7			
8			
9			
10			

	× 50	× 4	× 54
1			
2			
3			
4			
5			
6			
7			
8			
9			
10			

	× 70	× 2	× 72
1			
2			
3			
4			
5			
6			
7			
8			
9			
10			

	× 90	× 3	× 93
1			
2			
3			
4			
5			
6			
7			
8			
9			
10			

Now try this!

• **Use a similar method to answer these multiplications.**

$6 \times 27 =$ _____ $4 \times 65 =$ _____ $7 \times 42 =$ _____ $8 \times 84 =$ _____

$3 \times 79 =$ _____ $9 \times 58 =$ _____ $6 \times 73 =$ _____ $7 \times 91 =$ _____

Teachers' note For this activity the children should be familiar with multiplying a single digit by a multiple of 10 (for example, 4 × 70). Remind them that they can multiply the non-zero digits first and then multiply by 10 (4 × 7 = 28, 28 × 10 = 280). Point out that some facts in the tables may be found more easily using other methods (for example, 5 × 28 can be found by halving 10 × 28).

**Developing Numeracy
Mental Maths Year 6
© A & C BLACK**

Greetings cards

- **Find the area of each card. Write it in the table below.**

To find the area of a rectangle, multiply the length by the width.

A
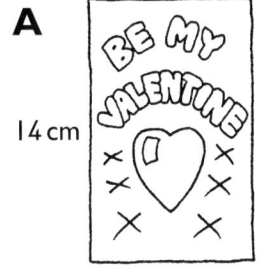
14 cm
9 cm

B
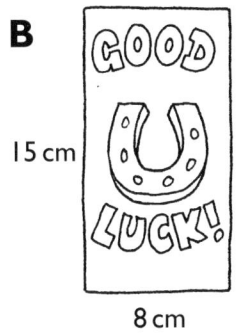
15 cm
8 cm

C

6 cm
19 cm

D

18 cm
9 cm

E
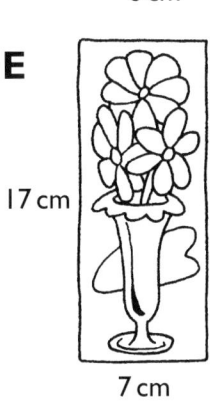
17 cm
7 cm

F
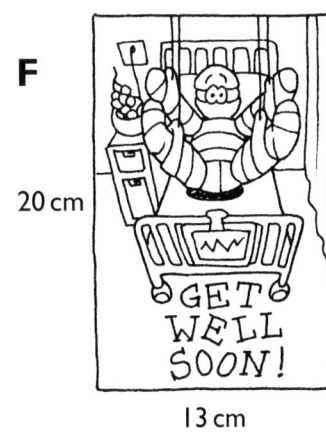
20 cm
13 cm

G
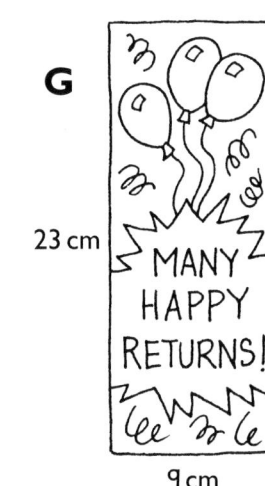
23 cm
9 cm

H

13 cm
7 cm

I
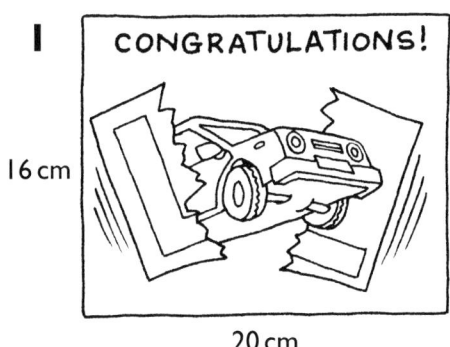
16 cm
20 cm

J

8 cm
16 cm

A	B	C	D	E	F	G	H	I	J
126 cm²									

- **List the cards in order of size. Start with the smallest.**

  H,

Now try this!

The sides of a card are whole numbers of centimetres. Its area is 96 cm² .

- **List all the possible lengths and widths of the card.**

Teachers' note This activity can be used to provide practice of partitioning a number before multiplying and then combining the parts. At the start of the lesson, revise tables facts and multiplying a multiple of 10 by a single-digit number (such as 20 × 7). Show how partitioning can be used to reach answers more quickly.

Developing Numeracy
Mental Maths Year 6
© A & C BLACK

Barmy brackets

The questions on each card are exactly the same except for the brackets.

Yes, brackets can totally change the answer to a question.

• **Answer the questions on each card.**

Do the parts in brackets first.

1.
$$8 \qquad 15$$
$9 - (2 \times 4) + (5 \times 3) = \underline{16}$

$(9 - 2) \times (4 + 5) \times 3 = \underline{\quad}$

$(9 - 2) \times 4 + (5 \times 3) = \underline{\quad}$

2.
$4 + (6 \times 3) + (2 \times 5) = \underline{\quad}$

$(4 + 6) \times (3 + 2) \times 5 = \underline{\quad}$

$(4 + 6) \times 3 + (2 \times 5) = \underline{\quad}$

3.
$6 + (3 \times 7) - (2 \times 2) = \underline{\quad}$

$(6 + 3) \times (7 - 2) \times 2 = \underline{\quad}$

$(6 + 3) \times 7 - (2 \times 2) = \underline{\quad}$

4.
$10 - (3 \times 2) - (2 \times 1) = \underline{\quad}$

$(10 - 3) \times (2 - 2) \times 1 = \underline{\quad}$

$(10 - 3) \times 2 - (2 \times 1) = \underline{\quad}$

5.
$1 + (6 \times 8) - (7 \times 2) = \underline{\quad}$

$(1 + 6) \times (8 - 7) \times 2 = \underline{\quad}$

$(1 + 6) \times 8 - (7 \times 2) = \underline{\quad}$

6.
$7 + (1 \times 4) + (5 \times 10) = \underline{\quad}$

$(7 + 1) \times (4 + 5) \times 10 = \underline{\quad}$

$(7 + 1) \times 4 + (5 \times 10) = \underline{\quad}$

7.
$7 + (5 \times 3) - (3 \times 7) = \underline{\quad}$

$(7 + 5) \times (3 - 3) \times 7 = \underline{\quad}$

$(7 + 5) \times 3 - (3 \times 7) = \underline{\quad}$

8.
$25 - (5 \times 4) - (3 \times 2) = \underline{\quad}$

$(25 - 5) \times (4 - 3) \times 2 = \underline{\quad}$

$(25 - 5) \times 4 - (3 \times 2) = \underline{\quad}$

Now try this!

• **Put brackets in these questions to make three different questions and answers.**

$8 + 2 \times 6 - 4 \times 3 = \underline{\quad}$

$8 + 2 \times 6 - 4 \times 3 = \underline{\quad}$

$8 + 2 \times 6 - 4 \times 3 = \underline{\quad}$

Teachers' note Ensure the children appreciate that if there are two or more sets of brackets in a question, it does not matter which they do first. Encourage them to write the answer to the parts in brackets above the question, to help them find the answer mentally. The children could write their own questions using more than two sets of brackets and find the answers.

Developing Numeracy Mental Maths Year 6 © A & C BLACK

Decimal digit game

- ## **Play this game with a partner.**

You need two sets of 0–9 digit cards and a copy of this sheet each.

☆ Shuffle the digit cards. Place them in a pile, face down.

☆ Turn over three cards. Both players should write the digits in the order which they think will give the largest answer.

☆ Work out the answer. Score a point if your answer is the larger.

☆ The winner is the player with the most points.

Example:

$2 \cdot 3 \times 6 = 13 \cdot 8$

$6 \cdot 2 \times 3 = 18 \cdot 6$

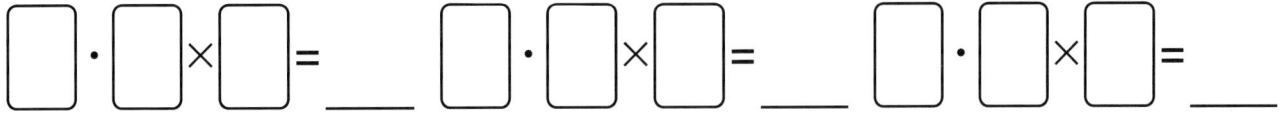

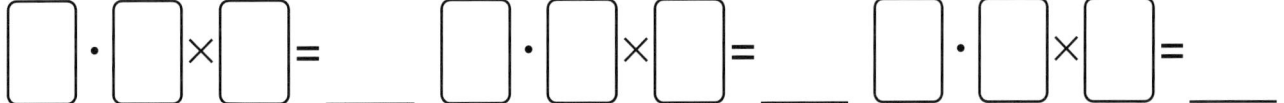

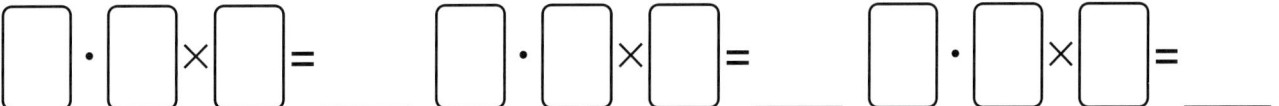

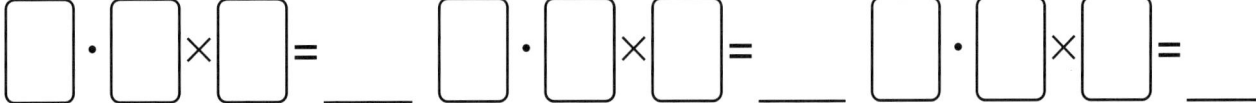

Now try this!

- ## **Shuffle the cards. Continue the game using this arrangement of digits.**

$0 \cdot \square\square \times \square = \underline{\quad}$ $0 \cdot \square\square \times \square = \underline{\quad}$

$0 \cdot \square\square \times \square = \underline{\quad}$ $0 \cdot \square\square \times \square = \underline{\quad}$

$0 \cdot \square\square \times \square = \underline{\quad}$ $0 \cdot \square\square \times \square = \underline{\quad}$

Teachers' note Discuss appropriate strategies for solving the questions. Partitioning can be used: for example, $3 \cdot 4 \times 6 = (3 \times 6) + (0 \cdot 4 \times 6) = 18 + 2 \cdot 4 = 20 \cdot 4$, or doubling and halving ($2 \cdot 5 \times 8 = 5 \times 4 = 20$). Remind the children that when multiplying a decimal by a single-digit number they can use their knowledge of tables and, if appropriate, divide the answer by 10 ($0 \cdot 4 \times 6 = 4 \times 6 \div 10 = 24 \div 10 = 2 \cdot 4$).

**Developing Numeracy
Mental Maths Year 6
© A & C BLACK**

37

Keep it simple

- **Follow the instructions to create** equivalent **fractions. Divide the numerator and the denominator by the same number in each step.**

> You can't exactly divide 1 and 3 by the same number (other than 1), so this fraction is in its **simplest form**.

Example:

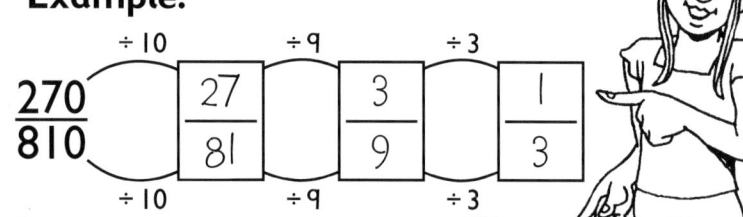

$$\frac{270}{810} \xrightarrow{\div 10} \boxed{\frac{27}{81}} \xrightarrow{\div 9} \boxed{\frac{3}{9}} \xrightarrow{\div 3} \boxed{\frac{1}{3}}$$

1. $\dfrac{18}{27}$ $\div 9$
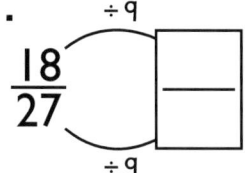

2. $\dfrac{56}{64}$ $\div 8$
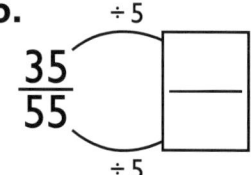

3. $\dfrac{160}{240}$ $\div 10$ $\div 8$
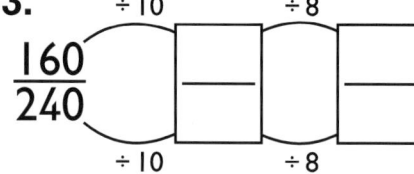

4. $\dfrac{49}{63}$ $\div 7$
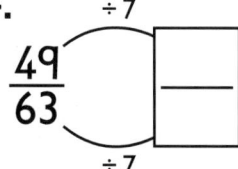

5. $\dfrac{35}{55}$ $\div 5$

6. $\dfrac{42}{140}$ $\div 2$ $\div 7$
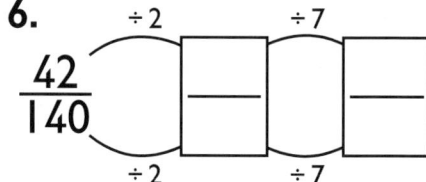

7. $\dfrac{48}{84}$ $\div 2$ $\div 6$
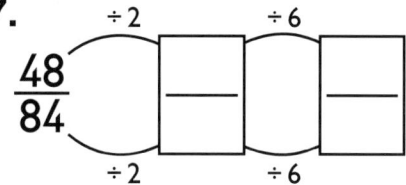

8. $\dfrac{300}{750}$ $\div 10$ $\div 5$ $\div 3$
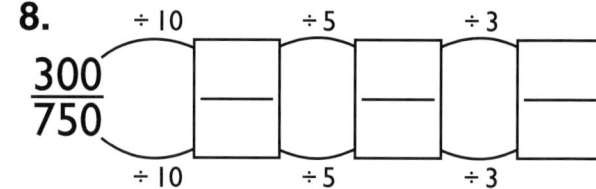

9. $\dfrac{36}{54}$ $\div 9$ $\div 2$
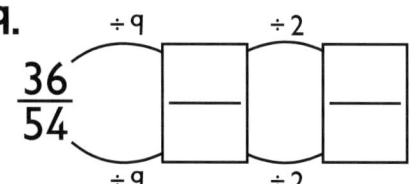

10. $\dfrac{96}{120}$ $\div 2$ $\div 6$ $\div 2$
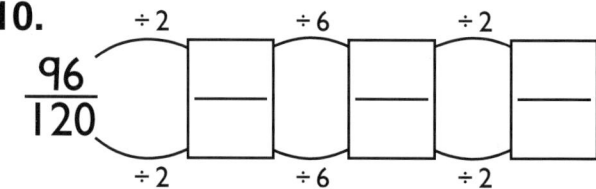

11. $\dfrac{1440}{1800}$ $\div 10$ $\div 2$ $\div 2$ $\div 3$ $\div 3$
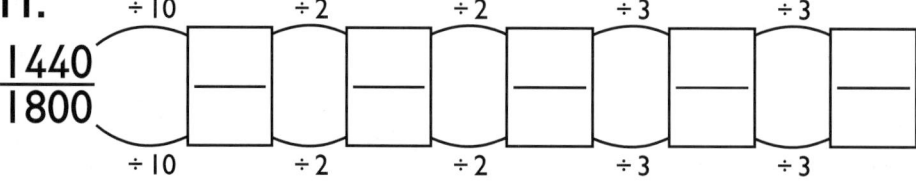

- **Change these fractions to their simplest form.**

$$\frac{180}{270} \qquad \frac{49}{56} \qquad \frac{2400}{3600} \qquad \frac{448}{640} \qquad \frac{616}{784} \qquad \frac{1248}{1920}$$

Teachers' note Ensure the children understand that the fractions are equivalent (the value does not change). Revise the terms 'numerator' and 'denominator'. Discuss different ways of reaching the simplest form: for example, for $\frac{16}{24}$ divide by 2, then by 2 and then by 2, or divide by 2 and then by 4, or divide by 8. Stress that the last way is quickest but relies on good tables knowledge.

Developing Numeracy Mental Maths Year 6 © A & C BLACK

Snail race

- ## Play this game with a partner.

☆ **You need** a dice and two small counters (snails).

☆ Take turns to roll the dice and move your snail. Work out the division and find the answer further along the path. Move your snail to this stone and wait until your next turn to roll the dice again.

☆ The first player to reach the plant pot is the winner.

start

2·5 — 33 ÷ 4	1·25 — 20 ÷ 8	$3\frac{1}{2}$ — 25 ÷ 2	$8\frac{1}{4}$ — 26 ÷ 5	$12\frac{1}{2}$ — 27 ÷ 4

2·5 — 49 ÷ 6

| 5·6 — 26 ÷ 3 | $4\frac{4}{9}$ — 29 ÷ 7 | $6\frac{3}{4}$ — 19 ÷ 5 | 9·5 — 28 ÷ 5 | $8\frac{1}{6}$ — 19 ÷ 2 |

5·2 — 40 ÷ 9

| 3·8 — 14 ÷ 4 | $4\frac{1}{7}$ — 28 ÷ 8 | $8\frac{2}{3}$ — 46 ÷ 5 | 3·5 — 24 ÷ 10 | 2·4 — 22 ÷ 3 |

$9\frac{1}{5}$ — 38 ÷ 9

| $4\frac{1}{4}$ — 34 ÷ 5 | $4\frac{6}{7}$ — 34 ÷ 8 | 0·9 — 3 ÷ 4 | $4\frac{2}{9}$ — 9 ÷ 10 | 4·5 — 34 ÷ 7 |

$7\frac{1}{3}$ — 27 ÷ 6

| $\frac{3}{4}$ — 30 ÷ 8 | 6·8 — 42 ÷ 4 | 3·75 — 41 ÷ 5 | $10\frac{1}{2}$ — 66 ÷ 9 | $3\frac{1}{3}$ — 49 ÷ 2 |

5·5 — 57 ÷ 7

| 8·25 — 59 ÷ 6 | 24·5 — 36 ÷ 5 | $9\frac{4}{5}$ — 43 ÷ 4 | $8\frac{1}{7}$ — 49 ÷ 5 | $8\frac{1}{5}$ — 39 ÷ 6 |

$7\frac{1}{3}$ — 33 ÷ 4

| 10·75 — 13 ÷ 10 | 6·5 — 17 ÷ 2 | 1·3 — 34 ÷ 4 | 7·2 — 51 ÷ 6 | $9\frac{5}{6}$ — 68 ÷ 8 |

8·5 **WIN!**

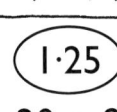
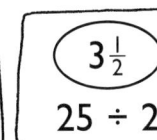
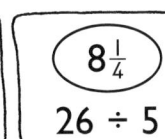
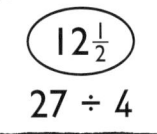

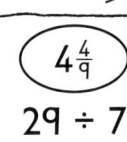
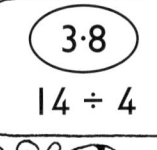
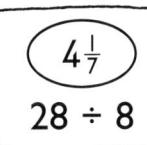
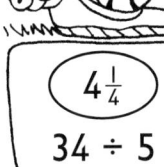
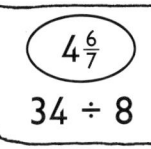
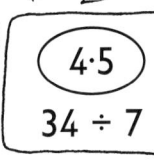
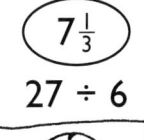
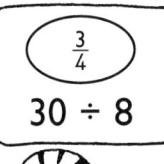
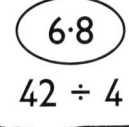

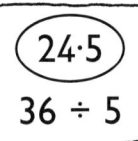
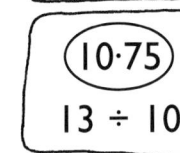
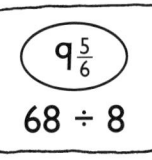

Teachers' note At the start of the lesson, remind the children that a remainder can be expressed as a fraction or as a decimal. Point out that some fraction remainders will need to be changed to their simplest form (for example, $3\frac{2}{8}$ becomes $3\frac{1}{4}$ and $5\frac{3}{6}$ becomes $5\frac{1}{2}$).

Developing Numeracy
Mental Maths Year 6
© A & C BLACK

39

Shopaholic

Hugo Spending always buys more than one of everything!

- **Write the cost of each item.**

1. 4 whistles cost £5.40 in total.

£ `1.35` each

2. 5 aprons cost £15.75 in total.

£ each

3. 4 combs cost £2.60 in total.

£ each

4. 6 scarves cost £10.50 in total.

£ each

5. 5 hats cost £6.75 in total.

£ each

6. 3 buckets cost £11.55 in total.

£ each

7. 3 mops cost £9.45 in total.

£ each

8. 5 irons cost £92.50 in total.

£ each

9. 4 staplers cost £11.80 in total.

£ each

10. 6 watches cost £22.50 in total.

£ each

11. 9 books cost £50.85 in total.

£ each

12. 8 T-shirts cost £33.20 in total.

£ each

- **Write the cost of each item.**

Now try this!

15 magazines cost £48.75 in total.

£ each

15 toothbrushes cost £34.50 in total.

£ each

Teachers' note The prices can be masked and changed to provide differentiation (for example, using only a whole number of pounds in each question). Discuss possible strategies for answering the questions and encourage the children to explain their thinking: for example, 'I know that four lots of £1 is £4, four lots of 30p is £1.20, and four lots of 5p is 20p. So, £5.40 ÷ 4 is £1.35.'

Developing Numeracy Mental Maths Year 6 © A & C BLACK

Not mushroom for fractions

Some of these mushrooms are edible and some are highly poisonous! The mushrooms with even answers are edible and the ones with odd answers are poisonous.

• **Answer the questions and tick the edible mushrooms.**

$\frac{2}{5}$ of 45 ✓   18

$\frac{4}{5}$ of 35

$\frac{5}{6}$ of 12

$\frac{3}{5}$ of 50

$\frac{2}{5}$ of 40

$\frac{3}{4}$ of 36

$\frac{3}{4}$ of 48

$\frac{3}{10}$ of 60

$\frac{3}{8}$ of 24

$\frac{7}{10}$ of 30

$\frac{4}{5}$ of 60

$\frac{5}{8}$ of 16

$\frac{4}{5}$ of 100

$\frac{5}{6}$ of 42

$\frac{2}{5}$ of 55

$\frac{5}{6}$ of 30

$\frac{7}{8}$ of 56

$\frac{5}{8}$ of 32

$\frac{9}{100}$ of 100

$\frac{3}{5}$ of 100

$\frac{5}{6}$ of 60

$\frac{3}{10}$ of 80

$\frac{4}{7}$ of 49

Now try this!

• **Write ten mushroom facts with the answer** 12 .

Thatshallot!

Teachers' note Stress to the children that they should *never* pick mushrooms. Discuss strategies for finding fractions of numbers, beginning with unit fractions. Ask: 'What is one-fifth of 15?' Show that this can be found by dividing by the denominator. Introduce non-unit fractions such as $\frac{3}{5}$, and demonstrate how to divide by the denominator and then multiply (for example, $\frac{1}{5}$ of 30 = 6, so $\frac{3}{5}$ of 30 = 6 × 3 = 18).

**Developing Numeracy
Mental Maths Year 6
© A & C BLACK**

Henry the eighth

• **Write the first ten** [multiples of 8] **on this counting stick.**
You could use doubling.

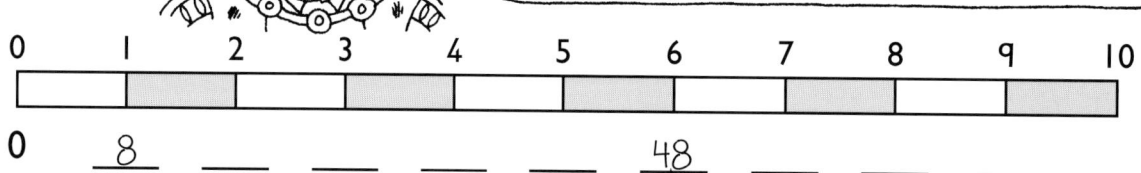

To multiply by 2, use 'double'.
To multiply by 4, use 'double, double'.
To multiply by 8, use 'double, double, double'.

0   1   2   3   4   5   6   7   8   9   10

0   __8__   ___   ___   ___   ___   __48__   ___   ___   ___   ___

double 6 = 12   double 12 = 24   double 24 = **48**

• **Change these improper fractions to mixed numbers.**

You can see from the counting stick that 48 eighths is the same as 6 whole ones. The first number is one-eighth more than that.

1. $\frac{49}{8}$ = __$6\frac{1}{8}$__   2. $\frac{25}{8}$ = _____   3. $\frac{43}{8}$ = _____   4. $\frac{45}{8}$ = _____

5. $\frac{57}{8}$ = _____   6. $\frac{33}{8}$ = _____   7. $\frac{77}{8}$ = _____   8. $\frac{71}{8}$ = _____

9. $\frac{28}{8}$ = _____   10. $\frac{55}{8}$ = _____   11. $\frac{47}{8}$ = _____   12. $\frac{79}{8}$ = _____

• **Now change these mixed numbers to improper fractions.**

The counting stick shows that 7 whole ones is the same as 56 eighths. The first number is three-eighths more than that.

13. $7\frac{3}{8}$ = __$\frac{59}{8}$__   14. $2\frac{1}{8}$ = _____   15. $7\frac{1}{8}$ = _____   16. $3\frac{3}{8}$ = _____

17. $4\frac{7}{8}$ = _____   18. $6\frac{2}{8}$ = _____   19. $9\frac{4}{8}$ = _____   20. $8\frac{5}{8}$ = _____

21. $2\frac{4}{8}$ = _____   22. $8\frac{7}{8}$ = _____   23. $5\frac{6}{8}$ = _____   24. $9\frac{6}{8}$ = _____

Now try this!

• **Answer these <u>without</u> using the counting stick.**

How many eighths is the same as:

(a) one and one-eighth? _____   **(b)** two and two-eighths? _____

(c) three and three-eighths? _____   **(d)** four and four-eighths? _____

(e) five and five-eighths? _____   **(f)** six and six-eighths? _____

(g) seven and seven-eighths? _____   **(h)** eight and eight-eighths? _____

• **Explain the pattern in your answers.**

Teachers' note At the start of the lesson, practise doubling numbers to find answers to '×2', '×4' and '×8' questions. When multiplying by 8, ask the children to check that they have doubled three times. As a further extension, the children could draw a counting stick to show multiples of 6 and explore converting between fractions and mixed numbers with a denominator of 6.

Developing Numeracy
Mental Maths Year 6
© A & C BLACK

Powerful percentages

- **Start at the central star. Follow the arrows to find percentages of** 240 **. Use the facts you complete to help you find other facts.**

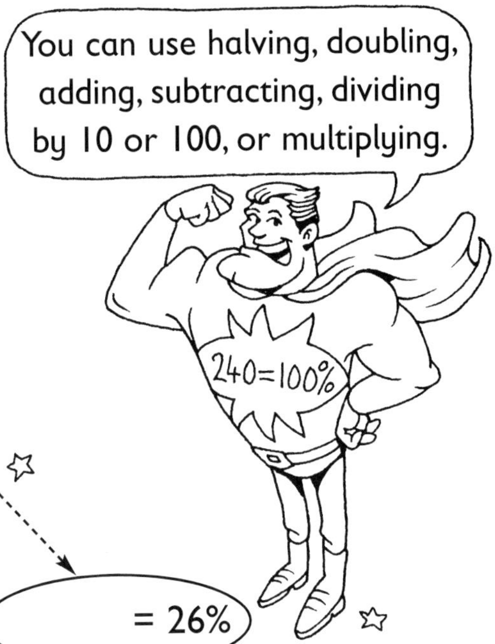

You can use halving, doubling, adding, subtracting, dividing by 10 or 100, or multiplying.

240 = 100%

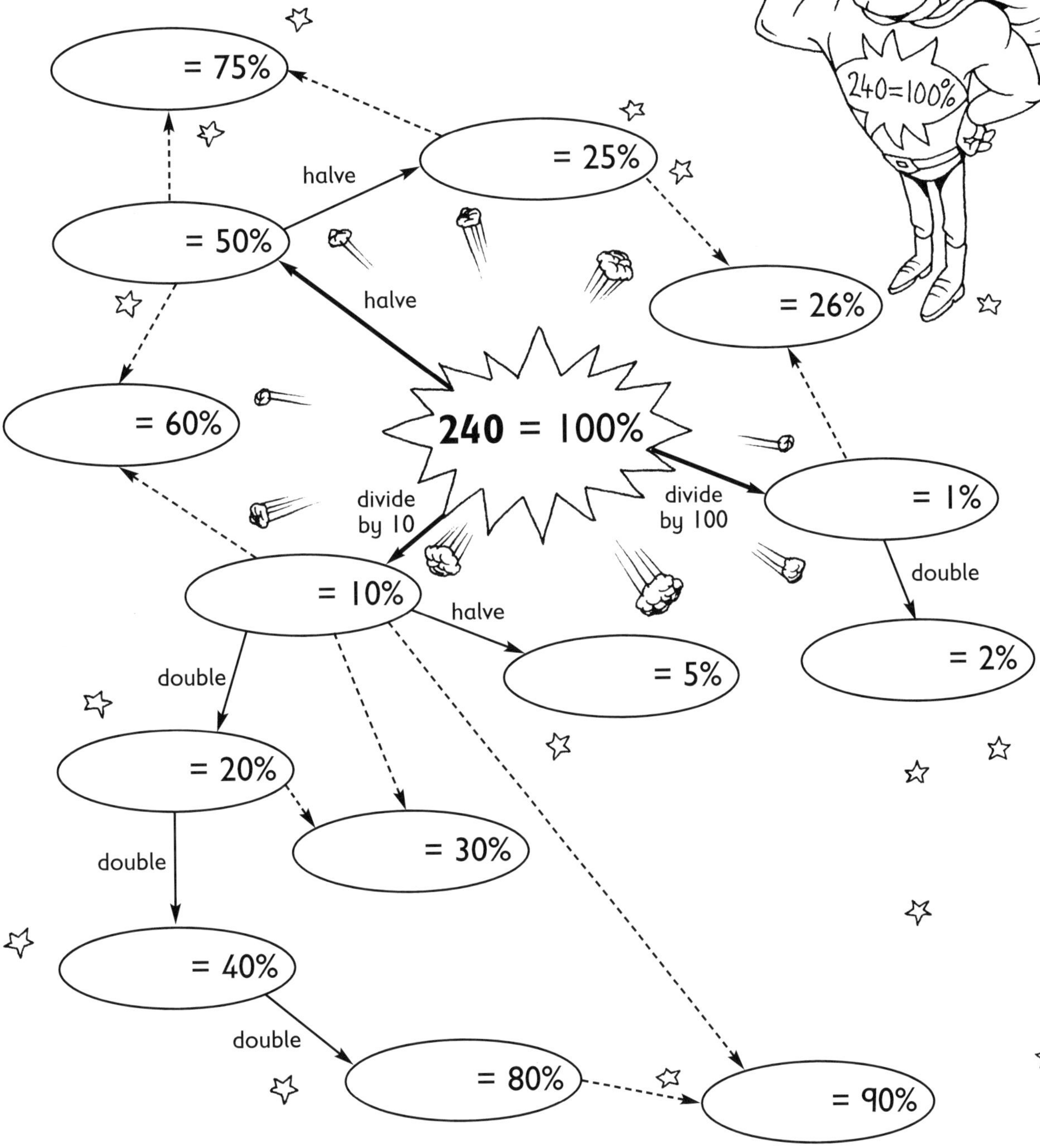

= 75%

= 25%

halve

= 50%

halve

= 26%

= 60%

240 = 100%

divide by 10

divide by 100

= 1%

double

= 10%

halve

= 5%

= 2%

double

= 20%

double

= 30%

double

= 40%

double

= 80%

= 90%

- **Add your own rings for other percentages you can find.**

- **Draw your own diagram with this fact in the central star:**

360 = 100%

Teachers' note At the start of the lesson, demonstrate how percentages of a number can be found mentally: for example, finding 10% by dividing by 10. Discuss how other percentages can be found: for example, multiplying 10% by 3 to find 30%. When completing the activity, tell the children to follow all the solid arrows first. Invite them to draw diagrams for other numbers such as 720, 750 and 960.

**Developing Numeracy
Mental Maths Year 6
© A & C BLACK**

43

Gift shop challenge

In this gift shop, you can pay in pounds or in euros.
The prices are shown in euros. The exchange rate
is £1 = €1.40 .

DVD €21.00

bookmark €1.05

bug finder €18.20

tea towel €3.50

T-shirt €16.80

cap €11.20

alien €4.20

pencil €0.70

dinosaur €12.60

book €15.40

mug €7.00

keyring €2.10

- **Each person spent all their money on one item. What did they buy?**

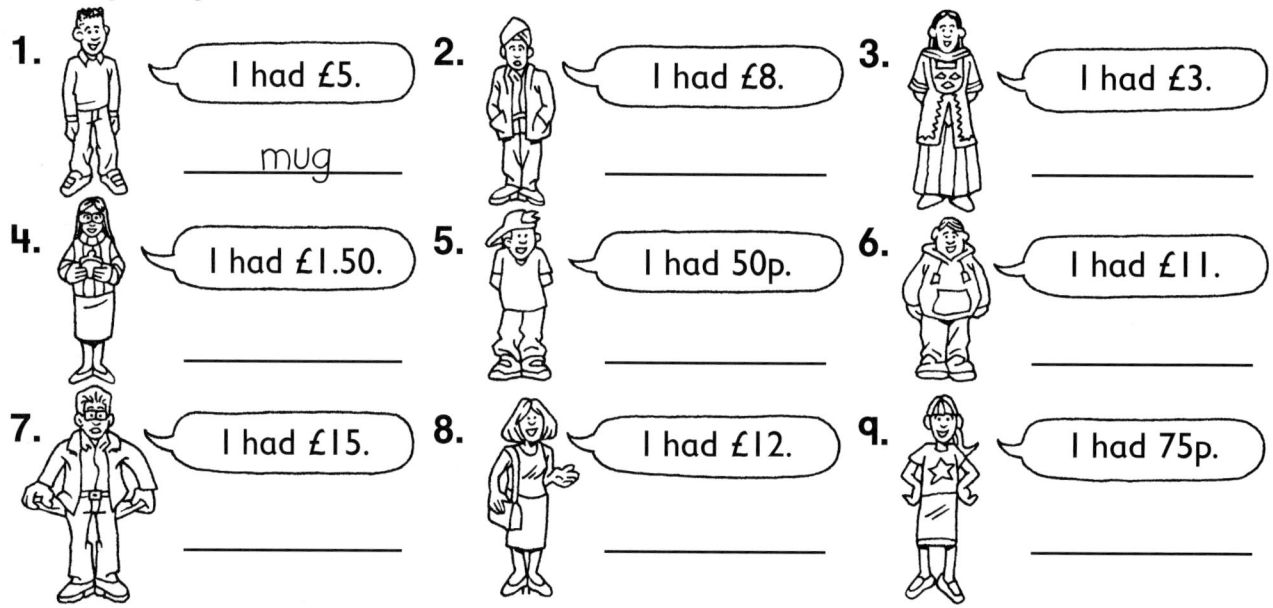

1. I had £5.

mug

2. I had £8.

3. I had £3.

4. I had £1.50.

5. I had 50p.

6. I had £11.

7. I had £15.

8. I had £12.

9. I had 75p.

- **Three items were not bought. Write their prices in pounds.**

- If £1 = €1.80 , how much money did each person have in euros?

Teachers' note Begin by explaining what is meant by 'exchange rate'. Discuss strategies for making these conversions mentally: for example, multiplying by 1·4 by multiplying by 1 and then by 0·4 and adding the two parts. As a further extension, encourage the children to make up different exchange rates and to convert into euros the number of pounds each person had.

**Developing Numeracy
Mental Maths Year 6
© A & C BLACK**

Stars and diamonds

• **Use the 'rule' at the top of each box to answer the questions.**

1. ☆ + ◇

8	6	→ 14
22	9	→
28	15	→
48	38	→

2. ☆ – ◇

54	6	→ 48
25	12	→
74	37	→
85	29	→

3. ☆ × ◇

3	8	→
6	9	→
7	8	→
8	12	→

4. (☆ × 2) + ◇

8	4	→ 20
12	7	→
6	9	→
7	15	→

5. (☆ × 3) ÷ ◇

8	4	→ 6
10	5	→
7	3	→
6	9	→

6. (☆ – ◇) × 2

20	6	→
26	4	→
50	12	→
25	9	→

 • **Work out the rules. Then fill in the last answer.**

☆　◇

36	9	→ 4
10	5	→ 2
24	4	→ 6
42	7	→

☆　◇

10	6	→ 14
25	12	→ 38
5	7	→ 3
13	21	→

☆　◇

3	8	→ 19
6	9	→ 24
14	10	→ 34
13	12	→

Teachers' note As a further extension, the children can create their own puzzles. Encourage them to use a range of operations and to include brackets if desired. They can either provide a rule for others to follow or give answers to a secret rule and encourage others to find it. These early ideas of algebra can be developed further by using *a* to stand for the star and *b* for the diamond.

Developing Numeracy
Mental Maths Year 6
© A & C BLACK

Passcode digits

To use a website, Sally needs to choose a passcode with five or more digits. She wants a passcode that she can turn into a maths fact, to help her remember it.

• Look at these passcodes and their maths facts.

253479	133930	25811	4503150

$25 \times 3 + 4 = 79$ $13 \times 3 - 9 = 30$ $2 \times 5 - (8 + 1) = 1$ $450 \div 3 = 150$

• Find maths facts to help Sally remember these passcodes.

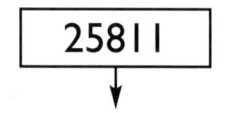 You can use any operations, and brackets.

1. 426613

$42 \div 6 + 6 = 13$

2. 84725

3. 306630

4. 271512

5. 79511

6. 150256

7. 745520

8. 384684

9. 89666

10. 444460

11. 67671

12. 352145

13. 250550

14. 4264100

15. 1234650

16. 1234677

• **Write a passcode with a maths fact that uses all four operations. Give it to a partner to solve.**

Teachers' note At the start of the lesson, provide examples of maths facts involving various operations and including brackets. Demonstrate how these could be written as passcodes: for example, $2 \times (3 + 4) = 14$ would make the passcode 23414. Remind the children that brackets show which parts of the calculation should be done first.

**Developing Numeracy
Mental Maths Year 6
© A & C BLACK**

Answers

p 6
Now try this!

£3200   £4140   £16420

p 7
1. 5·2	**2.** 53	**3.** 0·68
4. 570	**5.** 5·3	**6.** 980
7. 79	**8.** 852	**9.** 72
10. 0·46	**11.** 0·92	**12.** 0·46

Now try this!

4·5   6·2   8·95   11·32

6·76   3·22   30·05   65·17

p 8
Jump 1	Jump 2	Jump 3	Jump 4
⁻16	⁻15	⁻13	⁻11
8	12	12	20
⁻3	⁻9	⁻2	4
6	8	11	18
⁻2	⁻5	1	7
2	4	8	14
⁻1	⁻1	3	10
0	5	6	11

Now try this!

17	23	21
22	22	25
1	7	13
4	8	12

p 9
Next 12 square numbers:

81   100   121   144   169   196   225   256   289

324   361   400

Digital roots:

1	4	9	7	7	9	4	1	9	1
4	9	7	7	9	4	1	9	1	4

The sequence is repeated every 9 terms.

(240)   104 329   (10 241)   (501 203)

Now try this!

1   3   6   10   15   21   28   36   45

The sum of adjacent numbers forms the sequence of square numbers.

p 11
Now try this!

24 = 24   36 = 36   21 = 21

8 = 8   36 = 36 = 36 = 36

p 12
Possible answers:

1 + 4	4 + 4	4 + 9
9 – 4	25 – 16	49 – 36
1 + 49	16 – 1	25 – 4
25 + 25	64 – 49	121 – 100
4 × 4	4 × 16	4 × 9
4 × 25	16 × 9	25 × 9
16 ÷ 4	36 ÷ 9	100 ÷ 25
64 ÷ 16	100 ÷ 4	144 ÷ 4
81 ÷ 9	144 ÷ 16	36 ÷ 4

Now try this!

2, 5, 8, 10, 13, 17, 18, 20, 25, 26, 29

p 13
1. (a) 20	**(b)** 30
(c) 18	**(d)** 45
(e) 40	**(f)** 24
(g) 36	**(h)** 60
(i) 56	**(j)** 210

2. (a) 3 × 3 × 3	**(b)** 2 × 2 × 7
(c) 2 × 3 × 7	**(d)** 2 × 5 × 7
(e) 2 × 2 × 2 × 2	**(f)** 2 × 2 × 3 × 7
(g) 2 × 2 × 2 × 11	**(h)** 2 × 3 × 3 × 5
(i) 2 × 2 × 5 × 5	**(j)** 2 × 3 × 5 × 5

p 14
1. hexagon	**2.** pentagon	**3.** heptagon
4. triangle	**5.** trapezium	**6.** octagon

p 15
1. 200	800	**2.** 190	810
3. 170	830	**4.** 240	760
5. 410	590	**6.** 330	670
7. 90	910	**8.** 110	890
9. 390	610	**10.** 270	730

Now try this!

250   750   100   900

p 18
Board 1: 10, 12, 14, 15, 17, 19, 20, 21, 22, 24, 25, 26, 27, 28, 29, 30

Board 2: 8, 9, 10, all numbers from 12 to 30

Board 3: all multiples of 3 from 6 to 30

Board 4: all multiples of 2 from 4 to 30

p 19
CHIPS, EGGS, BEANS, TOAST, SHAKE

p 20
1. 71	**2.** 51	**3.** 23	
4. 62	**5.** 59	**6.** 15	**7.** 19
8. 84	**9.** 71	**10.** 78	**11.** 95

Now try this!

101

p 21
1. Increasing in 7s	**2.** Increasing in 11s
3. Decreasing in 4s	**4.** Increasing in 1s (row – 12; column + 13)

p 24

× 4
24	28	8
4	20	36
32	12	16

× 8
48	56	16
8	40	72
64	24	32

× 6
36	42	12
6	30	54
48	18	24

× 7
42	49	14
7	35	63
56	21	28

× 3
18	21	6
3	15	27
24	9	12

× 9
54	63	18
9	45	81
72	27	36

Now try this!

31	36	11
6	26	46
41	16	21

All the squares are magic.

p 27
1.
1 × 15 = 15	2 × 14 = 28	3 × 13 = 39
4 × 12 = 48	5 × 11 = 55	6 × 10 = 60
7 × 9 = 63	8 × 8 = 64 ✔	

2.
1 × 1 × 14 = 14	1 × 2 × 13 = 26	1 × 3 × 12 = 36
1 × 4 × 11 = 44	1 × 5 × 10 = 50	1 × 6 × 9 = 54
1 × 7 × 8 = 56	2 × 2 × 12 = 48	2 × 3 × 11 = 66
2 × 4 × 10 = 80	2 × 5 × 9 = 90	2 × 6 × 8 = 96
2 × 7 × 7 = 98	3 × 3 × 10 = 90	3 × 4 × 9 = 108
3 × 5 × 8 = 120	3 × 6 × 7 = 126	4 × 4 × 8 = 128
4 × 5 × 7 = 140	4 × 6 × 6 = 144	5 × 5 × 6 = 150 ✔

3.
1 × 1 × 1 × 13 = 13	1 × 1 × 2 × 12 = 24
1 × 1 × 3 × 11 = 33	1 × 1 × 4 × 10 = 40
1 × 1 × 5 × 9 = 45	1 × 1 × 6 × 8 = 48
1 × 1 × 7 × 7 = 49	1 × 2 × 2 × 11 = 44
1 × 2 × 3 × 10 = 60	1 × 2 × 4 × 9 = 72
1 × 2 × 5 × 8 = 80	1 × 2 × 6 × 7 = 84
2 × 2 × 2 × 10 = 80	2 × 2 × 3 × 9 = 108
2 × 2 × 4 × 8 = 128	2 × 2 × 5 × 7 = 140
2 × 2 × 6 × 6 = 144	3 × 3 × 1 × 9 = 81
3 × 3 × 2 × 8 = 144	3 × 3 × 3 × 7 = 189
3 × 3 × 4 × 6 = 216	3 × 3 × 5 × 5 = 225
4 × 4 × 1 × 7 = 112	4 × 4 × 2 × 6 = 192
4 × 4 × 3 × 5 = 240	4 × 4 × 4 × 4 = 256 ✔
5 × 5 × 5 × 1 = 125	5 × 5 × 4 × 2 = 200
6 × 5 × 4 × 1 = 120	6 × 5 × 3 × 2 = 180
6 × 6 × 3 × 1 = 108	7 × 4 × 3 × 2 = 168
7 × 5 × 3 × 1 = 105	8 × 4 × 3 × 1 = 96

¹1	0	²6	■	³1	2	⁴8	■	⁵2
6	■	⁶4	0	4	■	⁷5	0	0
■	⁸1	0	■	⁹9	¹⁰7	■	■	0
¹¹6	4	■	¹²1	■	¹³2	8	¹⁴8	
¹⁵1	5	■	¹⁶4	8	■	¹⁷4	0	
■	■	¹⁸1	4	■	¹⁹2	0	■	
²⁰1	0	3	■	²¹3	6	■	²²1	3
6	■	²³1	²⁴2	■	²⁵5	0	0	
²⁶9	0	■	²⁷3	4	■	■	²⁸9	9

p 30

1. 8000   2000   4700   5300
2. 12 800   3200   8600   7400
3. 21 200   5300   11 900   14 600
4. 20 600   5150   16 850   8900

Now try this!
$a = 2400$ $b = 1600$

p 31

1. 12   2. 15   3. 14   4. 54
5. 2·1   6. 6   7. 2·4   8. 8·1
9. 4·2   10. 6·3

Now try this!
49   72   99
9·9   2·1   3·6
5·6   1·8   3·5
7·7   5·4   8·1

p 33

1. 637
   650
   663
2. 735   3. 784   4. 1176   5. 1372
   750   800   1200   1400
   765   816   1224   1428
6. 2156   7. 1274   8. 2058   9. 1666
   2200   1300   2100   1700
   2244   1326   2142   1734

Now try this!
686   1122   882
1275   1764   1287
1782   2178   3232

p 34

17	66	28
34	132	56
51	198	84
68	264	112
85	330	140
102	396	168
119	462	196
136	528	224
153	594	252
170	660	280

54	72	93
108	144	186
162	216	279
216	288	372
270	360	465
324	432	558
378	504	651
432	576	744
486	648	837
540	720	930

Now try this!
162   260   294   672
237   522   438   637

p 35

A   B   C   D   E
126 cm²   120 cm²   114 cm²   162 cm²   119 cm²
F   G   H   I   J
260 cm²   207 cm²   91 cm²   320 cm²   128 cm²

H, C, E, B, A, J, D, G, F, I

Now try this!
96 cm × 1 cm   48 cm × 2 cm   32 cm × 3 cm   24 cm × 4 cm
16 cm × 6 cm   12 cm × 8 cm

p 36

1. 16   2. 32   3. 23   4. 2
   189   250   90   0
   43   40   59   12
5. 35   6. 61   7. 1   8. ⁻1
   14   720   0   40
   42   82   15   74

Now try this!
$8 + (2 \times 6) - (4 \times 3) = 8$
$(8 + 2) \times (6 - 4) \times 3 = 60$
$(8 + 2) \times 6 - (4 \times 3) = 48$

p 38

1. $\frac{2}{3}$ 2. $\frac{7}{8}$ 3. $\frac{2}{3}$
4. $\frac{7}{9}$ 5. $\frac{7}{11}$ 6. $\frac{3}{10}$
7. $\frac{4}{7}$ 8. $\frac{2}{5}$
9. $\frac{2}{3}$ 10. $\frac{4}{5}$
11. $\frac{4}{5}$

Now try this!
$\frac{2}{3}$ $\frac{7}{8}$ $\frac{2}{3}$ $\frac{7}{10}$ $\frac{11}{14}$ $\frac{13}{20}$

p 40

1. £1.35   2. £3.15   3. £0.65   4. £1.75
5. £1.35   6. £3.85   7. £3.15   8. £18.50
9. £2.95   10. £3.75   11. £5.65   12. £4.15

Now try this!
£3.25   £2.30

p 42

1. $6\frac{1}{8}$ 2. $3\frac{1}{8}$ 3. $5\frac{3}{8}$ 4. $5\frac{5}{8}$
5. $7\frac{1}{8}$ 6. $4\frac{1}{8}$ 7. $9\frac{5}{8}$ 8. $8\frac{7}{8}$
9. $3\frac{1}{2}$ 10. $6\frac{7}{8}$ 11. $5\frac{7}{8}$ 12. $9\frac{7}{8}$
13. $\frac{9}{8}$ 14. $\frac{17}{8}$ 15. $\frac{57}{8}$ 16. $\frac{27}{8}$
17. $\frac{39}{8}$ 18. $\frac{50}{8}$ 19. $\frac{76}{8}$ 20. $\frac{69}{8}$
21. $\frac{20}{8}$ 22. $\frac{71}{8}$ 23. $\frac{46}{8}$ 24. $\frac{78}{8}$

Now try this!
(a) 9   (b) 18   (c) 27   (d) 36
(e) 45   (f) 54   (g) 63   (h) 72

p 44

1. mug   2. cap   3. alien
4. keyring   5. pencil   6. book
7. DVD   8. T-shirt   9. bookmark
bug finder £13, tea towel £2.50, dinosaur £9

Now try this!
1. €9   2. €14.40   3. €5.40
4. €2.70   5. €0.90   6. €19.80
7. €27   8. €21.60   9. €1.35

p 45

1. 14   2. 48   3. 24
   31   13   54
   43   37   56
   86   56   96
4. 20   5. 6   6. 28
   31   6   44
   21   7   76
   29   2   32

Now try this!
☆ ÷ ◇   (2 × ☆) − ◇   ☆ + (2 × ◇)
last number 6   last number 5   last number 37

p 46

Possible answers:
1. 42 ÷ 6 + 6 = 13   2. 8 × 4 − 7 = 25   3. 30 ÷ 6 × 6 = 30
4. 27 − 15 = 12   5. 7 + 9 − 5 = 11   6. 150 ÷ 25 = 6
7. (7 − 4) × 5 + 5 = 20   8. 38 + 46 = 84   9. (8 × 9) − 66 = 6
10. (44 − 44) × 6 = 0   11. (6 × 7) ÷ 6 ÷ 7 = 1
12. 3 × 5 × (2 + 1) = 45   13. 250 ÷ 5 = 50
14. (4 × 26) − 4 = 100   15. (12 ÷ 3) + 46 = 50   16. 123 − 46 = 77